6/05

MACMILLAN

D0491581

Work Out

Biology

GCSE

The titles in this series

For GCSE examinations

Accounting	Modern World History
Biology	Human Biology
Business Studies	Core Maths Key Stage 4
Chemistry	Revise Mathematics to further
Computer Studies	level
English Key Stage 4	Physics
French (cassette and pack	Religious Studies
available)	Science
Geography	Social and Economic History
German (cassette and pack	Spanish (cassette and pack
available)	available)
	Statistics

For 'A' Level examinations

Accounting	Mathematics
Biology	Physics
Business Studies	Psychology
Chemistry	Sociology
Economics	Statistics
English	
French (cassette and pack	
available)	

Work Out

Biology

GCSE

O.F.G. Kilgour

MACMILLAN

First published 1986 by
THE MACMILLAN PRESS LTD
Houndmills, Basingstoke, Hampshire RG21 2XS
and London
Companies and representatives
throughout the world

ISBN 0–333–44005–6

A catalogue record for this book is available
from the British Library.

Printed and bound in Great Britain by
Biddles Ltd
Guildford and Kings Lynn

First edition reprinted (with corrections) 1986
Second edition 1987
10 9 8 7 6 5 4
01 00 99 98 97 96 95

Dedication:
To Mary and Dilys Kitty Brearley

£ 8-99
Dawson
PO15224

Contents

Acknowledgements

Every effort has been made to trace all the copyright holders, but if any have been inadvertently overlooked, the publishers will be pleased to make the necessary arrangements at the first opportunity.

Throughout the preparation of this book I have been grateful for the constructive criticism and advice of a long-established teacher of biology, Meinwen Parry, B.Sc., M.I.Biol., whose identification with the preparation of candidates for the combined CSE/GCE alternative biology syllabus of the Welsh Joint Education Committee since its inception has made an invaluable contribution to the work.

I am also appreciative of the assistance and encouragement given by my editor, Mary Waltham, B.Sc., M.I.Biol., of Macmillan Education.

I express my sincere thanks to Mrs V. C. Stirling for the professional and speedy preparation of the typescript. The author and publishers wish to thank Mr Malcolm Fraser of Chelsea College Centre for Science Education for permission to use Figs 1.5, 4.4, 14.6 and the photograph on page 40. The cover photograph, by Stephen Dalton (NHPA), shows *Apis mellifera* (honey-bee) landing on apple blossom.

Sincere appreciation is expressed to Lt. Commander Richard Frampton of the Marine Society and Seafarers Education Service, and to County Councillor Doris Griffiths OBE, for efficient and prompt support in obtaining the GCSE-approved syllabuses.

I appreciate the kindness of Dr Iolo Ap Gwynn of the Zoology Department, University College of Wales, Aberystwyth, for permission to use the electron micrograph showing the mitochondrion (page 50).

Grateful acknowledgement is made to the following examining boards for permission to reproduce their past examination questions: Associated Examining Board, University of London Entrance and Schools Examination Council, Oxford Local Examinations, Southern Universities Joint Board for School Examinations, Scottish Certificate of Education Examining Board, Secondary Education Authority, Western Australia, and Northern Ireland General Certificate of Education Examinations Council.

The University of London Entrance and School Examinations Council accepts no responsibility whatsoever for the accuracy or method in the answers given in this book to actual questions set by the London Board.

Acknowledgement is made to the Southern Universities' Joint Board for School Examinations for permission to use questions taken from their past papers but the Board is in no way responsible for answers that may be provided and they are solely the responsibility of the author.

The Associated Examining Board, the University of Oxford Delegacy of Local Examinations, the Northern Ireland Schools Examination Council and the Scottish Examination Board wish to point out that worked examples included in the text are entirely the responsibility of the author and have neither been provided nor approved by the Board.

Organisations Responsible for GCSE Examinations

In the United Kingdom, examinations are administered by the following organisations. Syllabuses and examination papers can be ordered from the addresses given here:

Northern Examining Association (NEA)

Joint Matriculation Board (JMB)
Publications available from:
John Sherratt & Son Ltd
78 Park Road, Altrincham
Cheshire WA14 5QQ

North Regional Examinations Board
Wheatfield Road, Westerhope
Newcastle upon Tyne NE5 5JZ

Yorkshire and Humberside Regional Examinations Board (YREB)
Scarsdale House
136 Derbyside Lane
Sheffield S8 8SE

Associated Lancashire Schools Examining Board
12 Harter Street
Manchester M1 6HL

North West Regional Examinations Board (NWREB)
Orbit House, Albert Street
Eccles, Manchester M30 0WL

Midland Examining Group (MEG)

University of Cambridge Local Examinations Syndicate (UCLES)
Syndicate Buildings, Hills Road
Cambridge CB1 2EU

Oxford and Cambridge Schools Examination Board (O & C)
10 Trumpington Street
Cambridge CB2 1QB

Southern Universities' Joint Board (SUJB)
Cotham Road
Bristol BS6 6DD

East Midlands Regional Examinations Board (EMREB)
Robins Wood House, Robins Wood Road
Aspley, Nottingham NG8 3NR

West Midlands Examinations Board (WMEB)
Norfolk House, Smallbrook
Queensway, Birmingham B5 4NJ

London and East Anglian Group (L & EAG)

University of London School Examinations Board (L)
University of London Publications Office
52 Gordon Square
London WC1E 6EE

London Regional Examining Board (LREB)
Lyon House
104 Wandsworth High Street
London SW18 4LF

East Anglian Examinations Board (EAEB)
The Lindens, Lexden Road
Colchester, Essex CO3 3RL

Southern Examining Group (SEG)

The Associated Examining Board (AEB)
Stag Hill House
Guildford
Surrey GU2 5XJ

University of Oxford Delegacy of
   Local Examinations (OLE)
Ewert Place, Banbury Road
Summertown, Oxford OX2 7BZ

Southern Regional Examinations
   Board (SREB)
Avondale House
33 Carlton Crescent
Southampton, Hants SO9 4YL

South-East Regional Examinations
   Board (SEREB)
Beloe House
2–10 Mount Ephraim Road
Royal Tunbridge Wells, Kent TN1 1EU

Scottish Examination Board (SEB)

Publications available from:
Robert Gibson and Sons (Glasgow) Ltd
17 Fitzroy Place, Glasgow G3 7SF

Welsh Joint Education Committee (WJEC)

245 Western Avenue
Cardiff CF5 2YX

Northern Ireland Schools Examinations
   Council (NISEC)

Examinations Office
Beechill House, Beechill Road
Belfast BT8 4RS

Introduction

This book has been written to meet the needs of students of biology preparing for external examinations at 16+, and it will be a useful aid to revision, and for practical and examination work throughout the course.

How to Use this Book

Each chapter is presented in the following way:

1.1 Theoretical work summary of the key information you should know and understand. Further detail is to be found in *Mastering Biology* by O. F. G. Kilgour (Macmillan Education).

1.2 A summary of the practical work you will be expected to have had experience of, and that will indicate what you can do.

1.3 Examination questions, which form the major part of the book, covering the wide range of question types used in examining candidates at this level. The questions are grouped into three main sections: objective, structured and free response.

Differentiation: the symbol ✳ in the margin indicates work required for the optional or extended examination and grades 'C' to 'A'.

Suggested answers to the free-response questions or essays follow straight on from these questions, and answers to the objective and structured questions are at the end of each chapter. Some short-answer free-response questions are not provided with answers, and require you to provide the answers by reading the text.

You should use this book during your course, when preparing answers for homework, and when revising for examinations. Use the objective and structured questions to test your grasp of each topic covered by a chapter.

Revision

All examining boards or associations issue a detailed syllabus and you should obtain a copy of this by writing to the appropriate address shown on pages ix and x. Read through the syllabus as you revise and make sure you understand each section properly. Do not study topics which are *not* included in your examination syllabus.

Throughout your course you should aim to work methodically. Plan your final revision in good time, work out a programme of revision and stick to it. Take each major topic, such as respiration, in turn, and revise it thoroughly. To test your grasp of each topic, attempt past examination papers under examination conditions; stick to the time allocated for each questions, and, of course, do not use your notes to help you. In biology it is important to be able to draw and label diagrams accurately and you will need to practise doing this as you revise. Try to simplify the diagrams that you draw so that you can complete them quickly and clearly.

The GCSE Examination

(a) Written Examinations

Written external examinations follow different patterns and use different types of question. Obtain past question papers from the appropriate examining board (see pages ix and x) to find out the pattern and type of question, the number of questions to be answered, and whether certain sections are composed of compulsory questions. (See Table I.)

The written examinations for the GCSE examination in the United Kingdom usually consist of *two* main papers concerned with finding out what you know and understand of the *theoretical principles* of biology, as listed in the subject content of the syllabuses issued by different examining groups.

I. The *basic or common level* written examination is usually a single paper which is *compulsory* for all candidates and is usually of 2 hours' duration. The results of this examination award grades from G to C. Some examining bodies may have *two* written examinations for this basic or common level first examination.

II. The *optional extended level* written examination is usually a single paper of about 1½ hours duration. The results of this examination award grades between E and A.

The following are the main types of written examination questions.

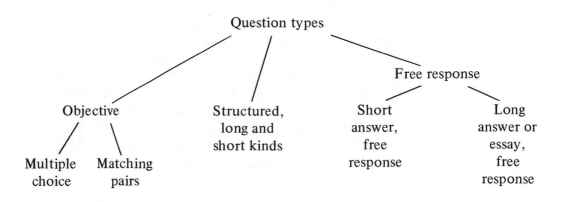

(i) *Objective Questions*

This type of question is mainly of the kind that has four, or sometimes five, answers or responses, one of which is correct. They require very careful reading and thought, and certainly cannot be answered by guessing.

Objective questions will be found in some first basic or common level examinations and are of two different types, multiple choice and matching pairs, and mainly consist of a *statement* or *stem* and a number of *answers* or *options*, one of which, the *key*, is correct; the others are incorrect *distractors*.

1. *Multiple choice* questions are the most common kind of objective question in the basic level first examination, with up to *five* answers or options, only *one* of which is correct.

Example

Which one of the following parts of the eye is sensitive to light? (This is the question *stem*.)
Answers or *options*:

a. retina   c. iris
b. lens   d. optic nerve

The *correct* answer or *key* is 'a', retina. The others are incorrect or distractors.

2. *Matching pairs* are classification-type objective questions and are less common in the first basic level examination. This type of question consists of either four or five terms, statements or facts which must be *matched* correctly with their respective partners in four or five questions. (See Chapter 16.)

Example

Questions 1 to 4
    a. Saprophytism
    b. Competition
    c. Parasitism
    d. Predation.

From the above list of biological terms, choose the one that describes:
    Q.1   yeast growing on rotten fruit
    Q.2   birds eating worms
    Q.3   fungus growing on live plant leaves
    Q.4   animal body decaying in soil.

The correct matching is as follows:
    Q.1   matches with a.
    Q.2   matches with d.
    Q.3   matches with c.
    Q.4   matches with a.

Note: The option 'b' does not match with any question. Candidates should *carefully check* matching pairs for this possibility and of a 'match' being used more than once, as 'a' is in Q.1 and Q.4.

(ii)   *Structured Questions*

These questions can be either short or long and may take a long time to read, but once they are understood they are among the quickest and most direct questions to answer in the examination paper. Structured questions can be based on experimental data, and you may need to draw a graph, or they may be based on diagrams, photographs or written statements. The questions usually require *short* answers. The length of your answer can be gauged roughly by the space left for your answer or by the number of marks awarded, which is shown on the paper. This type of question can occur in either basic or optional examination paper.

(iii)   *Free-response Short-answer Questions*

The length of the written answer expected is shown in the answer book of the examination paper as one, two or more lines. Sometimes they are completion-type questions requiring the insertion of missing words or terms. This type of question occurs in the first, basic examination paper.

(iv) *Free-response Long-answer or Essay Questions*

This type of question is always found in the second, optional or extended examination paper. These are mainly unstructured questions, requiring the careful planning of an answer by the candidate. Consequently, worked examples are given in this book to assist the reader with a wide range of problems. Remember that in essay-type answers there are many ways of presenting acceptable answers and candidates can achieve the same overall mark by many different answers; only one version is shown here. Essay questions can be *direct* questions on certain topics, or *mixed* or composite questions on more than one biological topic.

Various *terms* are used in questions, which means that it is essential to understand their meaning. Most of the terms relate to the type of answer required.

(i) Concise, short, descriptive answers are generally expected for questions that include: 'state', 'outline', 'define', 'state and explain', or include the word 'briefly' followed by a term.

(ii) Longer written answers are generally expected for questions that include: 'describe', 'discuss', 'explain', 'give an account', or 'comment', or which may include the word 'fully', 'in detail', or 'detailed'.

(iii) *Diagrams* or *graphs* are expected to be drawn in answer to questions that include: 'illustrate', 'make', 'sketch', 'draw', 'give annotated diagrams', or 'an illustrated account'.

(iv) *Tabular* answers are generally expected to questions that ask you to: 'distinguish', 'enumerate', 'tabulate', 'compare', or 'contrast and compare'.

(v) *Practical questions* often include the following, asking how you would: 'demonstrate', 'find', 'estimate', 'measure' or 'calculate'.

(v) *Diagrams (see also page 283)*

Diagrams are *line* drawings, i.e. they are made up of continuous lines. They are *not* artistic sketches, made up of broken lines. They should be drawn in pencil, large and clear, and labelled fully, and the completed diagram should be given a title and if possible an indication of its scale of magnification. *No* shading, colouring or three-dimensional constructions are needed. Draw diagrams whenever possible and *do not* write a written description of what is shown in your diagram. Certain diagrams will be given in an examination for you to *label* or *explain*. See, for example, Figure 3.2.

(vi) *In the Examination*

(i) Read the *instructions* on the paper.
(ii) Read the *questions* carefully, think, then answer.
(iii) Time flies so keep to a time allocation for each question.

(b) Practical Skills (See Chapter 15)

Biology, like most science subjects, is a practical subject that involves visual observation by the eye, the use of the hands in various *manipulative* (*manus*: meaning 'hand', Latin) handling techniques using apparatus, instruments, tools, substances or specimens. Hearing, smell and tasting are employed to a lesser extent in scientific experimentation.

Certain practical skills are compulsorily examined or assessed by teachers during normal classroom activity some time before the written examinations. This course assessment of practical skills can earn between 20% and 30% of the total marks for the whole GCSE examination, leaving you to earn between 70% and 80% in the written examination.

The course assessment of practical skills is *compulsory* for all candidates to GCSE taking either basic or extended-level written examinations.

External candidates not in full-time education will be required to attempt a practical examination of about 1½ hours duration, based mainly on the practical skills described in Chapter 15. Details must be obtained by individual external candidates by writing to the relevant examining group.

Table I GCSE types of examination question and marks awarded (1988)

Examination group	First written compulsory basic: number of questions				Second written optional extended: number of questions	
	Multiple choice	*Matching pairs*	*Short to medium structured*	*Longer structured*	*Long or medium structured*	*Free-response essay*
SEG	—	—	20	2	6	—
MEG	50	—	15	—	3	2
NEA	24	16	13	—	15	—
L & EAG	—	9	14	4	2	2
NISEC	13	12	18	—	6	—
WJEC	—	—	22	1[a]	15	2

[a] Long, free response

Question type	Marks awarded (estimated)
Short structured	Up to 6
Medium structured	Between 7 and 13
Long structured	Between 14 and 21
Long, free response	Up to 20
Multiple choice	1 each
Matching pairs	1 each

Note. This is an *estimated* plan based on information given in specimen examination papers issued in 1986. It may be subject to revision by the examining groups.

1 Basic Bioscience

1.1 Theoretical Work Summary

(a) Bioscience

1. *Biology* is the knowledge or science of life.
2. *Scientific method*:
   - (i) involves *observations* and repeated *experiment*;
   - (ii) *theories* and *hypotheses* are ideas based on observations;
   - (iii) *laws* are widely accepted and fully proved theories.
3. *Qualitative observations* deal with the biological *variables* of living organisms, such as colour, smell, shape or chemical composition.
4. *Quantitative observations* deal with the *quantity* or variables which can be *measured*.
5. *Control experiments* are identical to the investigation experiment in all respects except *one*, the factor being investigated, which is kept constant.
6. *Experimental information* is recorded as data, and this information is interpreted in various ways.

(b) Biological Measurement

1. *Prefixes* are used as the first part of the term for units of measurement.
   Mega-, M, one million times, 10^6
   Kilo-, k, one thousand times, 10^3
   Milli-, m, one-thousandth, 10^{-3}
   Micro-, μ, one-millionth, 10^{-6}
   Nano-, n, one-thousand-millionth, 10^{-9}
2. *Metre*, symbol m, unit of length. (See page 271.)

   100 centimetres, cm   1 000 000 micrometres, μm

   1 metre, m

   1000 millimetres, mm   1 000 000 000 nanometres, nm

   Area measured in square metres, m^2, or hectares, $10\,000$ m^2 (2.47 acres). (See page 272).
3. *Litre*, symbol l, unot of volume.
   One litre = 1000 millilitres, or dm^3, or 1000 cubic centimetres, cm^3. (See page 273.)
4. *Gram*, symbol g, unit of measurement of mass or weight.
   One kilogram, kg = 1000 grams, g.
   One gram = 1000 milligrams, mg = $1\,000\,000$ micrograms, μg.
5. *Concentration* is the *amount* of a substance in a known amount of another substance, e.g. mass in 1000 cm^3 or 1 litre. Also can be *percentage* weight/ weight or weight/volume. (See page 279).

Solution, a mixture formed when a *solute* dissolves in a *solvent*.

Solute + solvent = solution

Molarity is a method of expressing concentration as moles of substance in 1 kg, or 1 dm^{-3}, of solvent or water. One molar sucrose contains one mole or 342 g in 1000 g water; 0.1 molar sucrose is one-tenth of 342 = 34.2 g in 1000 g water, or 0.1 mole dm^{-3}.

6. *Surface area to weight or volume ratio*
The relationship between surface area and volume is illustrated in Fig. 1.1.

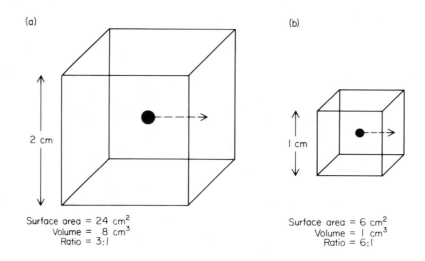

Figure 1.1   Relationship between surface area and volume. As the body *increases* in size the amount of surface compared with volume *decreases*.

$$\frac{\text{Surface area}}{\text{volume}} = \text{ratio}$$

$$\frac{\text{Surface area}}{\text{weight}} = \text{ratio}$$

As plants, animals and cube models become *larger*, their surface area becomes *smaller* relative to their volume or weight. Similarly the distance from the centre to the body surface *increases* with increasing body size.

(c) Biometry

This is the application of mathematics to the study of living organisms. (See also section 15.6(c).)

1. *Graphs* are diagrams showing the relationship between one changing or fixed quantity and another. 'Units' or quantities *must* be indicated on each axis, e.g. time in seconds, time (s), mass in kilograms, mass (kg), etc. (See page 287).
2. *Bar charts* are used to display statistics or numerical facts concerning large numbers of related living organisms.
3. *Pie diagrams* are circular diagrams for comparing resemblances or differences between living organisms or the composition of biological materials. Each quantity occupies a sector of the circle equal to a certain number of *degrees*. (See page 286.)

(d) Biophysics

This is the application of physics to biology.
1. *Energy forms* are interconvertible and are the means to do work.
    (a) *Chemical* energy is stored in chemical compounds within cells, e.g. adenosine triphosphate (ATP).
    (b) *Heat* energy is always produced when energy is changed into different forms.
    (c) *Kinetic* energy is the energy of movement or motion.
    (d) *Radiant* energy is the energy of light and is composed of the electromagnetic spectrum radiating from the sun (see Fig. 1.2).

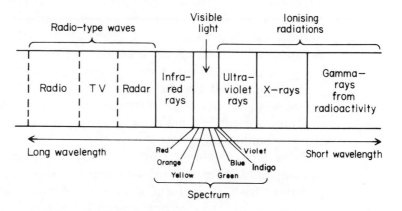

Figure 1.2 The electromagnetic spectrum.

2. *Energy measurement* is in the basic unit of the *joule*, J.

    One kilojoule, kJ = 1000 joule
    One megajoule, MJ = 1000 kJ = 1 000 000 J

3. *Temperature* is a relative measure of hotness or coldness on the Celsius scale, °C. Boiling point of water = 100°C. Freezing point of water = 0°C. Average human body temperature = 37°C.
4. *Pressure* is a force exerted by solids, liquids or gases, in or on the bodies of living organisms. The unit of pressure is the pascal (Pa). (See page 274).

    One kilopascal, kPa = 1000 Pa. Centimetres, and millimetres, are alternative units for pressure measurement: see page 274.

5. *Diffusion* is the movement of particles from a region of high concentration to one of a lower concentration, until the concentration is uniform or homogeneous in the solution. The difference in concentration is called the *concentration gradient*; when no concentration gradient exists, the solution is in *equilibrium*.
6. *Osmosis* is the passive diffusion of water, or a solvent, through a selectively permeable *membrane* (cell membrane, dialysis or visking tubing) into a more concentrated solution from a region of *lower* concentration, tending to equalise the concentrations on both sides of the membrane.

(e) Biochemistry

This is the study of chemical processes and substances occurring in living organisms.
    1. *Chemical elements* are the 100 different substances, ranging from hydrogen, symbol H, to fermium, symbol Fm, composing living and non-living matter.

2. *Atoms* are the smallest component parts of a chemical element. Each atom consists of a central nucleus (protons and neutrons) surrounded by an outer orbit of electrons. The atom is electrically *neutral* with an equal number of electrons and protons.

3. *Isotopes* are chemical elements with atoms of different *mass* due to different numbers of neutrons in the nucleus. A chemical element may have several isotopes, shown as follows for carbon: ^{12}C, ^{13}C or ^{14}C; or ^{16}O, ^{17}O, ^{18}O, for oxygen.

4. *Radioactivity*, in which atoms disintegrate to produce alpha, beta, or gamma *radiation*, detected by geiger counters or photographic film, is a property of certain radioactive isotopes, e.g. ^{35}S, ^{32}P and ^{14}C. Isotopes of ^{18}O and ^{15}N are *not* radioactive and are detected by means of the *mass spectrometer*.

5. *Tracers* are either radioactive or heavy non-radioactive isotopes used to trace or follow chemical changes in living organisms.

6. *Chemical compounds* are composed of two or more elements joined together by chemical *bonds*. The smallest part of a chemical compound able to exist on its own is called a *molecule*.

7. *Chemical or molecular formulae* are used to show the *kind* and *number* of atoms present in a molecule, e.g. CO_2 is the chemical formula for a molecule of carbon dioxide.

8. *Inorganic compounds* are chemical substances excluding complex carbon compounds of non-living or *mineral* origin. They are taken into living organisms mainly in the form of atoms of elements or groups of atoms of different elements with an *electrical charge*, called *ions*.

9. *Organic compounds* are mainly complex compounds of carbon, formed in living organisms; they are mainly electrically neutral, have a high energy value, and form a very high percentage of the *dry* component of living organisms.

10. *Metabolism* is the sum total of all chemical changes that occur in living organisms. They include *anabolism* or synthesis of organic compounds, and *catabolism* or breakdown of organic compounds.

11. *Metabolic pathways* are the long, complex, stepwise connections, along which metabolism of both kinds proceeds.

	ANABOLISM Metabolic pathway	
Simple molecules + ENERGY	ENZYMES ——————————→ ←——————————	Large complex organic molecules
	Metabolic pathway CATABOLISM	

12. A summary of the molecular composition and important structural features of organic compounds of biological importance is given in Table 1.1.

13. *Enzymes* are functional proteins which increase the *rate* of chemical reactions in the metabolism of living organisms. *Cofactors* such as metal ions (K^+, Mg^{2+}) and vitamins of the B group are *coenzymes* needed for the activity of certain enzymes. There are two types of enzyme:
    (a) intracellular, found within the cell cytoplasm;
    (b) extracellular, which are secreted from the cell as digestive juices.

Enzyme properties

(i) *Specific* in action in converting one kind of substance (*substrate*) into another.

(ii) *Heat sensitive*, working best at an *optimum temperature* (30°C plants, 37°C human body); above this temperature, enzymes are destroyed by *denaturation* and *coagulation*. Below the optimum temperature, a rise of

Table 1.1 Composition and structural features of organic compounds

Class of organic compound	Chemical elements present	Structural features and functional group
Alcohols, names end in -ol, e.g. ethanol, propanetriol (glycerol); also the complex compounds cholesterol and cholecalciferol	C, H, O	Functional group OH or hydroxyl
Acids, names end in acid, e.g. ethanoic (acetic) acid, hydroxypropanoic (lactic) acid, many alkanoic (fatty) acids, and vitamin C (ascorbic acid)	C, H, O	Functional group COOH or carboxyl group
Lipids, all edible oils or fats, also waxes	C, H, O	Triglycerides formed from different alkanoic (fatty) acids and propanetriol (glycerol)
Carbohydrates 1. *Monosaccharides*, names end in -ose:	C, H, O	*Formula* ($C_1:H_2:O_1$ is the ratio of elements)
glucose and fructose	C, H, O	$C_6H_{12}O_6$
ribose and deoxyribose	C, H, O	$C_5H_{10}O_5$, $C_5H_{10}O_4$
trioses	C, H, O	$C_3H_6O_3$
2. *Disaccharides*, names end in -ose:		*Formed from*:
maltose	C, H, O	two glucose units, $C_{12}H_{22}O_{11}$
sucrose	C, H, O	glucose and fructose
lactose	C, H, O	glucose and galactose
3. *Polysaccharides*, names mainly end in -ose. Usually not sweet; water insoluble:		
cellulose	C, H, O	up to 3000 glucose units, $C_{7.2}:H_1:O_8$
amylose – starch	C, H, O	up to 600 glucose units
glycogen – animal starch	C, H, O	up to 1000 glucose units
Organic bases, names mainly end in -ine: adenine (A), thymine (T), guanine (G), cytosine (C), and uracil (U)	C, H, O and N	All contain nitrogen amine group
Nucleic acids ribonucleic acid, RNA deoxyribonucleic acid, DNA	C, H, O, N and P*	Three components in large molecule: (i) ribose or deoxyribose *sugar* (ii) organic *bases*, A, C, G, T, or U (iii) *phosphate* group; all joined together in a polynucleotide chain
Amino acids, about 20 different kinds with names mainly ending in -ine, e.g. alanine; -phan, e.g. tryptophan; or acid, e.g. glutamic acid	C, H, O and N (methionine, cystine and cysteine also contain S)	Contain carboxyl *acid*, and amine or *base* group
Proteins, names end in: -in, e.g. keratin, trypsin, albumin and myosin; -en, e.g. collagen; -ase, enzymes such as amylase and urease	C, H, O, N, S and P	Composed of many amino acid units, joined by *peptide* links to form long *polypeptide* chains. Proteins are either: (i) *structural* in muscle, skin and bone (ii) *functional* in enzymes and hormones

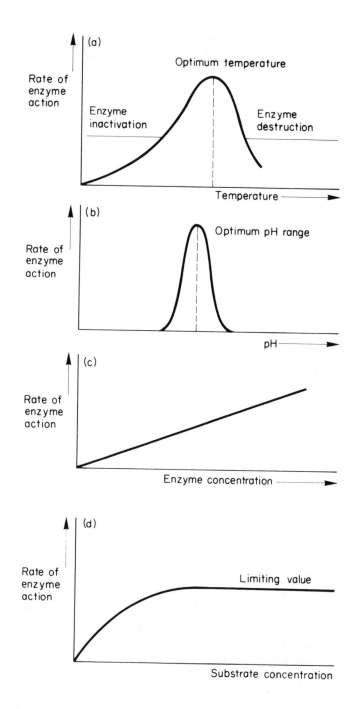

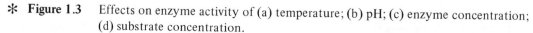

✳ Figure 1.3 Effects on enzyme activity of (a) temperature; (b) pH; (c) enzyme concentration; (d) substrate concentration.

10°C doubles the reaction rate: this is called the *temperature coefficient*. (See Fig. 1.3(a).)

(iii) *pH sensitive*. The pH scale is a numerical scale showing *acidity*, pH 1 to 7; *alkalinity*, over pH 7 to 14, or neutral, pH 7. Enzymes work within a narrow range of optimum pH. (See Fig. 1.3(b).)

(iv) *Sensitive to poisons*, e.g. lead, arsenic, cyanide and DDT.

(v) *Enzyme concentration*. As the concentration of enzyme increases, the rate of enzyme action also increases. Enzymes are not destroyed by the chemical change and can be used repeatedly. (See Fig. 1.3(c).)

(vi) *Substrate concentration*. As the amount of substance on which an enzyme acts (the *substrate*) increases, the rate of enzyme action increases to a maximum limiting value. (See Fig. 1.3(d).)

14. *Hydrolysis* is an important biochemical reaction in which a substance reacts with water, mainly aided by enzymes.

$$Sucrose + water \longrightarrow glucose + fructose$$

HYDROLYSIS

$$C_{12}H_{22}O_{11} + H_2O \longrightarrow 2C_6H_{12}O_6$$

15. *Condensation* is the reverse biochemical reaction to hydrolysis, in that smaller molecules combine to form larger molecules and eliminate water.

CONDENSATION

$$Many\ amino\ acids \longrightarrow protein + water$$

Organic compounds, e.g. *proteins, polysaccharides* and *nucleic acids*, are synthesised by enzyme action involving condensation.

1.2 Practical Work

(a) Diffusion

Figures 1.4(a) and (b) show the apparatus used to demonstrate diffusion in a gel solution and in air.

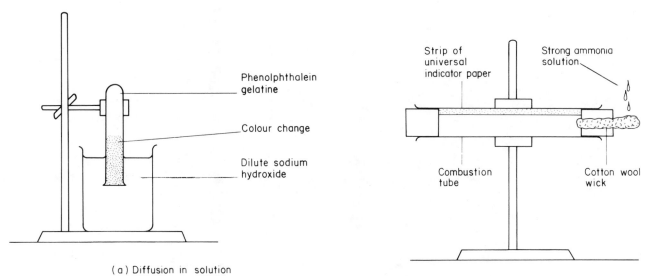

(a) Diffusion in solution

(b) Diffusion in air

Figure 1.4 Apparatus used to demonstrate diffusion in a liquid and a gas.

(b) Osmosis

1. Figure 1.5 is a variation of the apparatus used to demonstrate osmosis. Visking tube is a semi-permeable material also called dialysis tubing; an alternative material is cellophane.
2. (i) Prepare molar and M/2, M/4, M/8, M/16, or 0.5, 0.25, 0.125 and 0.062 mole dm^{-3} respectively, sucrose solutions (see 1.1(b) 5).

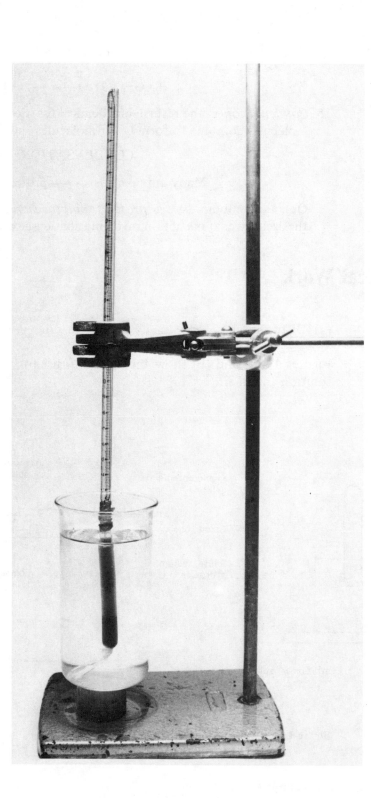

Figure 1.5 Apparatus used to demonstrate osmosis.

(ii) Use a cork borer to cut narrow potato cylinders, or discs, each weighing 10 g.

(iii) Place one in each sugar solution for 24 hours.

(iv) Dry off surplus liquid (using blotting paper) and reweigh.

(v) Determine which sample shows no change in weight. (Change in length can be measured instead of weight.)

(c) pH Determination

Universal Indicator test papers are used, first with a *broad range*, and then with a *narrow range*. In conjunction with a colour chart they give an approximate pH value for animal fluids or plant juices, and in soil testing.

(d) Identification of Chemical Substances

There are various tests that can be carried out to identify chemical substances of biological importance. These are summarised in Table 1.2.

Table 1.2   Tests for the identification of chemical substances

Chemical substance	Tests
Water	Turns *white* anhydrous copper sulphate blue or cobalt chloride paper from blue to pink
Carbon dioxide	(i) Clear calcium hydroxide solution (lime water) turns cloudy (ii) Hydrogen carbonate indicator (thymol blue and cresol red) changes from a red-purple colour to orange-yellow (iii) Soda lime and sodium or potassium hydroxide *absorb* CO_2
Oxygen	No test. Alkaline pyrogallol solution absorbs oxygen
Reducing sugars: glucose, fructose, maltose and ribose (sucrose is a **non-reducing** sugar)	Benedict's is better than Fehling's solution. Heat the blue solution and sugar together to boiling in a water bath. A green to brown or red precipitate forms. 'Clinistix' is a specific reagent strip test for *glucose*
Starch	Pale brown potassium iodide and iodine solution turns a blue-black colour
Lipids	(i) A warm sample pressed on paper makes a *translucent* permanent grease mark (ii) Mix oil sample with 3 cm^3 ethanol and add 3 cm^3 water: a milky *emulsion* forms
Proteins	(i) 'Albustix' reagent strip: colour changes from yellow to green (ii) 'Biuret' test: add few drops sodium hydroxide and 1% copper II sulphate: violet colour appears
Vitamin C (ascorbic acid)	Decolourises a blue solution of dichlorophenol indophenol (DCPIP)

(e) Conditions Affecting Enzyme Action

Catalase is an enzyme which breaks down hydrogen peroxide into oxygen (seen as gas bubbles) and water; it is found in *fresh* plant and animal material.

1. *Temperature*
   (i) Place samples of fresh liver or apple for 5 minutes in test tubes of water at 0, 10, 20, 30, 40, 50 and 60°C and boiling, 100°C.
   (ii) Remove the sample, add to a solution of dilute hydrogen peroxide and compare the rate at which oxygen is released.
2. *pH*
   (i) Place the liver or apple samples in solutions of different pH (1 to 14) for 5 minutes.
   (ii) Remove and test as in (ii) above.

1.3 Examination Work

(a) Multiple-choice Objective Questions

1. What is the ratio of carbon atoms to hydrogen and oxygen atoms in glucose?
   (a) $1:2:1$
   (b) $1:2:2$
   (c) $1:3:1$
   (d) $2:1:2$

2. Starch, a reserve food of a seed, is converted to glucose by the process of:
   (a) photosynthesis
   (b) translocation
   (c) hydrolysis
   (d) respiration

3. Which of the following reagents could be used to detect the presence of glucose in the germinating seed?
   (a) iodine solution
   (b) sodium or potassium hydroxide solution
   (c) alkaline pyrogallol
   (d) Benedict's or Fehling's solution

4. One of the following statements is incorrect concerning enzymes:
   (a) they are destroyed by high temperatures
   (b) they speed up some chemical changes
   (c) they are sensitive to pH change
   (d) they are only used in digestion

5. The Fehling's or Benedict's test is a biochemical test for:
   (a) starch
   (b) reducing sugars
   (c) glycogen
   (d) cellulose

6. Simple proteins are composed of:
   (a) fatty acids
   (b) amino acids
   (c) monosaccharides
   (d) glycerol

7. A concentrated solution of a black dye was prepared by dissolving a large amount of dye in a small amount of water. A small drop of this dye solution was carefully placed at the bottom of a large beaker of water. The beaker was kept still. After several days:
   (a) the drop of dye remained at the beaker bottom because it was denser than water
   (b) the drop floated to the water surface
   (c) the black dye gradually spread through the water
   (d) the black dye formed small solid particles at the beaker bottom

8. Enzymes are essential for biochemical reactions because they:
   (a) regulate the speed of the reaction
   (b) work best at temperatures over $50°C$
   (c) are not sensitive to pH changes
   (d) are used up in the chemical change

9. The complete hydrolysis of a starch amylose molecule produces products called:
   (a) alkanoic (fatty) acids
   (b) disaccharides
   (c) monosaccharides
   (d) polysaccharides

10. When an enzyme produced in the buccal cavity of human beings passes into the stomach it ceases to function because:
    (a) foods need different enzymes
    (b) this enzyme is affected by pH change
    (c) temperatures vary between the mouth and stomach
    (d) the digestion products have been absorbed

11. Which of the following reagents is used to test the urine of a diabetic patient?
    (a) 'Biuret test'
    (b) iodine solution
    (c) Benedict's solution
    (d) ammonium hydroxide

12. The main action of an enzyme in a biological system is primarily to:
    (a) produce energy
    (b) lower energy production
    (c) catalyse a specific reaction
    (d) produce water by condensation

13. The process by which the level of a sugar solution inside a tube attached to a differentially permeable membrane separated from pure water is seen to rise is called:
    (a) absorption
    (b) capillarity
    (c) osmosis
    (d) plasmolysis

14. The process which causes pure water in a beaker separated from a sugar and red ink solution contained in a differentially permeable plastic bag to turn red is called:
    (a) absorption
    (b) diffusion
    (c) osmosis
    (d) plasmolysis

15. Starch is changed into glucose by the chemical process of:
    (a) hydrolysis
    (b) condensation
    (c) neutralisation
    (d) oxidation

(b) Structured Questions

Question 1.1

(a) Name a naturally occurring substance that gives an orange colour as a result of Benedict's (or Fehling's) test. **(1 mark)**

(b) Three preparations of saliva were made and kept for 5 minutes at $40^\circ C$. Then a small sample of each was tested with iodine solution and another sample was tested with Benedict's solution. The resulting colours are recorded in the table below:

Preparation	Iodine test	Benedict's test
Saliva	yellow/pale brown	blue
Saliva and starch	yellow/pale brown	orange
Starch	blue/black	blue

Briefly describe what you can deduce from the tabulated results of this experiment. **(3 marks)**

(c) What would you expect the results of these two tests to be if the saliva is boiled then cooled before mixing with the starch?
  (i) boiled saliva and starch: iodine test **(1 mark)**
  (ii) boiled saliva and starch: Benedict's test **(1 mark)**

(d) If the above reaction was just complete in 5 minutes, how long would it have taken if the reaction had been carried out at $30^\circ C$? **(1 mark)**

(OLE)

* Question 1.2

In an investigation into the effect of temperature on enzyme action, the enzyme urease and the indicator bromo-thymol blue were mixed and then maintained at $35^\circ C$ for 5 minutes. (Bromo-thymol blue is yellow in neutral solution and blue in alkaline solution.) A pinch of urea was then added and within a few seconds an intense blue coloration had formed. The time taken for the blue colour to form was noted and the experiment was repeated at other temperatures. When urease and indicator (or urea and indicator) only were mixed together, no blue colour formed.

The results of the experiment are shown in the table below.

Temperature ($^\circ C$)	Time taken for blue colour to form (seconds)
0	93
15	23
35	6
45	18
55	35

(a) Plot these results on graph paper.

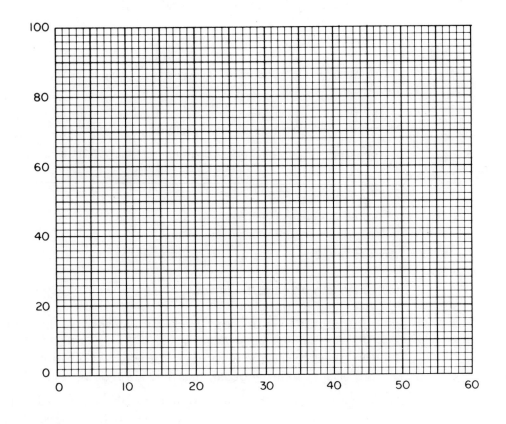

(b) What is an enzyme?

(c) Suggest why the indicator changes colour when urease is added to urea.

(d) At which temperature was urease most active?

(e) Suggest a reason for the length of time taken for the blue colour to form: (i) at $0°C$; (ii) at $55°C$. **(10 marks)**

(AEB, 1982)

(c) Free-response-type Questions

(i) *Short-answer Free-response Questions*

The time allowed for these is 5–10 minutes each.

1. Which two classes of substances found in foods contain all of the chemical elements carbon, oxygen and hydrogen?

2. Why is starch unable to pass through a dialysis membrane whereas water can?

3. State the meaning of diffusion.

4. Give an example of a food which is rich in protein and show how you would demonstrate this fact by means of a test.

The time allowed for these is about 30 minutes each.

* **Question 1.3**

(a) What are the characteristics of enzymes and what is their importance to living organisms?
(7 marks)

(b) Describe an experiment which demonstrates that the efficiency of a *named* enzyme is affected by an increase of temperature. (6 marks)

(c) Name two factors other than temperature which could affect enzyme efficiency.
(2 marks)

(d) What is a lipid fat? (5 marks)

(e) What is a carbohydrate? (5 marks)

Answer to Q1.3

(a) Enzyme *characteristics* include:
   (i) they are proteins
   (ii) they catalyse or speed up chemical changes of metabolism
   (iii) they are *unchanged* chemically in the reaction
   (iv) they are specific in action on usually *one* substrate
   Importance to living organisms: they are found in *every* living cell.

(b) Named enzyme: *amylases*, found in saliva, break down *starch* into disaccharides and a small amount of glucose, by hydrolysis.

Experiment

METHOD

1. Prepare 1% starch solution and place about 5 cm^3 in five test tubes.
2. Collect fresh saliva, firstly rinsing mouth free of food debris. Chew on a clean rubber band to encourage salivation. Place saliva in four test tubes.
3. Place starch and saliva solution in test tubes in a beaker of (a) ice; (b) water at room temperature; (c) water at 37–40°C, i.e. about blood heat; (d) boiling water.
4. Control: this consists of starch solution alone in warm water at about 37–40°C *without* the saliva test tube.
5. Allow test tubes to remain in the ice or water baths for 2 minutes to attain the bath temperature. Then add the saliva solution to the starch solution — except in control. Start the timer clock or stopwatch.

MEASUREMENT

Every minute, test one drop of the saliva/starch mixtures removed with a dropper with iodine solution. Note the time it takes for any one mixture to turn pale brown or remain blue-black — showing starch is changed into disaccharides.

RESULTS

The fastest colour change is seen in the mixture at the optimum temperature, 37–40°C, followed by that at room temperature of about 20°C. The others including the control produce an unchanged blue-black, showing starch is not changed by enzymes at a temperature of 0°C (ice mixture), or 100°C (boiling water), or where no enzyme is present in the control.

(c) Other factors affecting enzyme efficiency:
  (i) pH range, e.g. amylase requires pH 7 (neutral)
  (ii) *concentration* of the enzyme or the starch solution

(d) A lipid *fat* is a solid organic compound at room temperature consisting of alkanoic (fatty) acids and propanetriol (glycerol). (Lipid *oils* are liquid at room temperature.)

(e) A *carbohydrate* can be any one organic compound composed of one or more sugar or monosaccharide units, and also include disaccharides and polysaccharides.
  (*Note: Since* both *the organic compound groups lipids and carbohydrates contain carbon, hydrogen and oxygen, it is not enough to answer the question by saying so.*)

* **Question 1.4**

(a) State the meaning of diffusion and concentration gradient.   **(8 marks)**

(b) Describe a simple experiment to demonstrate diffusion in water.   **(6 marks)**

(c) What is osmosis?   **(4 marks)**

(d) Describe an experiment to demonstrate osmosis.   **(7 marks)**

Answer to Q1.4

This is a *direct* question.

(a) *Since the question includes the word 'state', a concise answer is expected.*

  Diffusion is the movement of substances or molecules from a region of high concentration to a region of lower concentration in liquids and gases until the substance is uniformly distributed.
  Concentration gradient is the difference in concentration of the substance which causes diffusion. The greater the concentration gradient the greater the *rate* of diffusion.
  At *equilibrium* there is no concentration gradient and the rate of diffusion is nil or zero.

(b) *Since the question refers to* water, *the following diagram is drawn.*

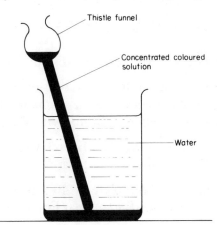

Thistle funnel

Concentrated coloured solution

Water

15

METHOD

1. Prepare a concentrated coloured solution of copper sulphate or a water-soluble food-colour dye.
2. Use a thistle funnel with its end drawn to a point, or a long-stemmed dropper to place a small amount of coloured concentrated solution as a thin layer at the beaker bottom.

RESULTS

When the beaker and its contents are allowed to stand in a cool place without disturbance for several days, the coloured substance is seen to *diffuse* or spread through the water.

(c) *Osmosis* is the *diffusion* of water (or solvent) through a *selectively permeable membrane*, from a region of high water (or solvent) concentration to a region of lower concentration.

(*Note*: It is necessary to explain *selectively permeable membrane* as: a porous membrane of dialysis or cellophane tubing which allows *small* particles or molecules, e.g. water and glucose, to pass easily and rapidly, whereas larger particles or molecules, e.g. starch and proteins, are restricted.)

(d) *A description of the apparatus and a diagram like Fig. 1.5 are needed here. Mention the important* precautions *to be taken*:
   (i) Presoak the dialysis tubing in plain tap water.
   (ii) Tie a firm knot in the dialysis tubing end, then fill with glucose solution, and tie the other end tightly to the glass capillary tube by means of thread.
   (iii) Wash the outside of the dialysis tube with clean tap water before immersing in the beaker of water.

1.4 Self-test Answers to Objective and Structured Questions

Answers to Multiple-choice Objective Questions

1. a   2. c   3. d   4. d   5. b   6. b   7. c   8. a   9. c   10. b   11. c   12. c   13. c   14. b   15. a

Answers to Structured Question 1.1

(a) Glucose, fructose, maltose or ribose

(b) Saliva alone has no action on iodine or Benedict's solutions; saliva and starch must have undergone a change to produce glucose; starch alone reacts with iodine solution but not with Benedict's solution.

(c) (i) boiled saliva and starch will cause iodine solution to turn blue or blue-black
    (ii) boiled saliva and starch will have no effect on Benedict's solution

(d) Reaction complete at 40°C in 5 minutes, therefore at 30°C it will take *twice* as long, or 10 minutes. (For every 10°C increase the rate is doubled.)

Answers to Structured Question 1.2

(a) Graph form: see diagram.

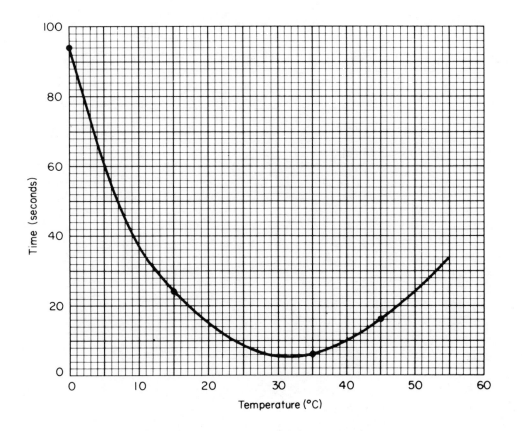

(b) Proteins which act as catalysts within cells.

(c) The enzyme urease must decompose urea into an *alkaline* product:

Urea ⟶ ammonia + carbon dioxide
urease

[*This knowledge is not needed to answer the question.*]

(d) From the graph the optimum temperature is about 30–34°C or on average 32°C.

(e) (i) at 0°C the enzyme is inactivated
(ii) at 55°C the enzyme is being destroyed by denaturation

Theme I

Diversity of Living Organisms

2 Living Organisms

2.1 Theoretical Work Summary

(a) Properties of Living Organisms

(i) *Nutrition* is the process of feeding and the subsequent digestion and assimilation of food nutrients.

(ii) *Growth* is the product of *synthesis* (the use of certain simple food nutrients in making an organism) and *cell division*.

(iii) *Respiration* is the release of energy from certain nutrients.

(iv) *Irritability* (sensitivity, responsiveness) is the organism's detection of and response to changes in the environment.

(v) *Movement* of the whole organism (locomotion) or part occurs internally and externally, and can be a consequence of irritability.

(vi) *Homeostasis* is the maintenance of balanced or constant conditions within the organism. It can involve metabolic waste removal, and control of water, nutrients and temperature.

(vii) *Reproduction* is the formation of new organisms, and the maintenance of *life*.

(b) Classification of Living Organisms

1. *Natural classification* is the arrangement of living and extinct organisms in a system of graded ranks, based on structural and functional similarities. The classification can also show evolutionary relationships from a common ancestor.

2. *Binomial (two-name) nomenclature* is the method of giving every organism two Latin names, first the *genus* name, written with a capital initial letter, followed by the *species* name; both names are underlined or printed in italics.

Common name	Generic name	Specific name
Housefly	*Musca*	*domestica*
Meadow fescue	*Festuca*	*pratensis*

There are two *kingdoms*:

(a) Animal kingdom of 940 000 species or 65% of all organisms.

(b) Plant kingdom of 440 000 species or 30% of all organisms.

There is also a *subkingdom*, protista, the microscopic plants (protophyta) and microscopic animals (protozoa) forming 5% of all organisms.

3. *Examples of classification* are provided in Table 2.1.

Table 2.1 Classification of the meadow buttercup *Ranunculus acris*, and the dog,
Canis familiaris

	Organisms	
Rank group	Meadow buttercup	Domestic dog
1. Kingdom	Plants	Animals
2. Division*	Vascular plants (Tracheophyta)	—
2. Phylum*	—	Chordata
3. Class	Angiospermae	Mammalia
4. Subclass	Dicotyledons	Eutheria
5. Order	Archichlamydeae	Carnivora
6. Family	Ranunculaceae	Canidae
7. Genus	*Ranunculus*	*Canis*
8. Species	*R. acris*	*C. familiaris*

*The *phylum* is used in animal classification whereas
the *division* is used in plant classification.

4. Table 2.2 summarises the differences between plants and animals.

Table 2.2 A comparison of the plant and animal kingdoms

Animal kingdom	Plant kingdom
65% of species in the world	30% of species in the world
1. No *plastids* in cells	1. Mainly green-coloured plastids or *chloroplasts* in cells — except fungi
2. Complex rapid responses	2. Slow simple response
3. Cannot *synthesise* organic compounds from inorganic substances. Organic food obtained from plants and animals, i.e. *heterotrophic*	3. *Synthesis* of organic compounds occurs using simple inorganic substances, i.e. *autotrophic* by photosynthesis or chemosynthesis
4. Food *consumers* and *decomposers*. *Detrivores* feed on detritus (mainly dead plant material)	4. Food *producers* except fungi
5. Metabolism forms *toxic* waste, removed by *excretion*	5. Metabolic waste seldom toxic, and not excreted
6. No cellulose in body structure	6. Cellulose cell walls

* 5. *Plant classification* can be summarised as shown below; examples of members of the plant kingdom are illustrated in Fig. 2.1.

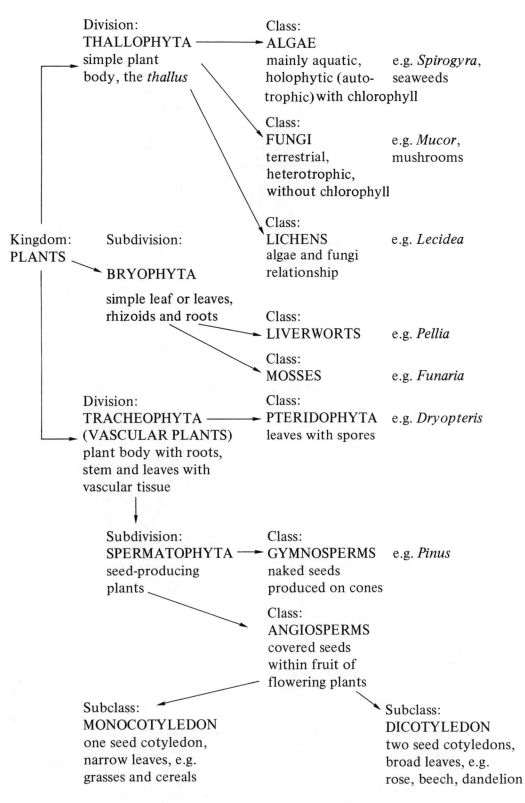

Kingdom: PLANTS

Division: THALLOPHYTA simple plant body, the *thallus*

Class: ALGAE mainly aquatic, holophytic (auto- trophic) with chlorophyll — e.g. *Spirogyra*, seaweeds

Class: FUNGI terrestrial, heterotrophic, without chlorophyll — e.g. *Mucor*, mushrooms

Class: LICHENS algae and fungi relationship — e.g. *Lecidea*

Subdivision: BRYOPHYTA simple leaf or leaves, rhizoids and roots

Class: LIVERWORTS — e.g. *Pellia*

Class: MOSSES — e.g. *Funaria*

Division: TRACHEOPHYTA (VASCULAR PLANTS) plant body with roots, stem and leaves with vascular tissue

Class: PTERIDOPHYTA leaves with spores — e.g. *Dryopteris*

Subdivision: SPERMATOPHYTA seed-producing plants

Class: GYMNOSPERMS naked seeds produced on cones — e.g. *Pinus*

Class: ANGIOSPERMS covered seeds within fruit of flowering plants

Subclass: MONOCOTYLEDON one seed cotyledon, narrow leaves, e.g. grasses and cereals

Subclass: DICOTYLEDON two seed cotyledons, broad leaves, e.g. rose, beech, dandelion

(a) Schematic diagram showing the internal structure of a bacterial cell – decomposer

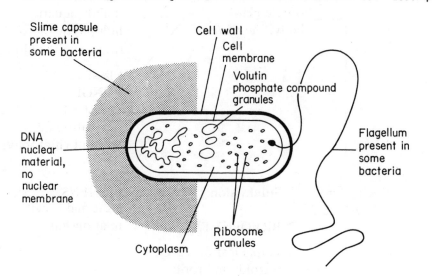

Slime capsule
present in
some bacteria

Cell wall

Cell
membrane

Volutin
phosphate compound
granules

Flagellum
present in
some
bacteria

DNA
nuclear
material,
no
nuclear
membrane

Ribosome
granules

Cytoplasm

EUKARYOTES

(b) Structure of the blue-green mould, *Penicillium* sp. – decomposer

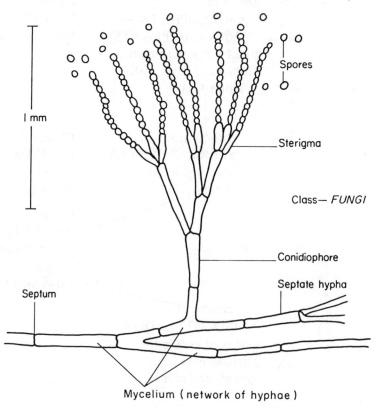

1 mm

Spores

Sterigma

Class— *FUNGI*

Conidiophore

Septate hypha

Septum

Mycelium (network of hyphae)

Figure 2.1   Examples of the plant kingdom.

(c) Structure of the algae *Chlamydomonas* sp. and *Spirogyra* sp. – producers

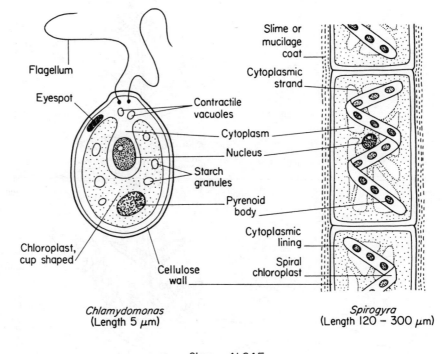

Chlamydomonas
(Length 5 μm)

Spirogyra
(Length 120 – 300 μm)

Class— *ALGAE*

Subdivision— *BRYOPHYTA*

(d) Examples of bryophytes: a liverwort, *Pellia* sp., and a moss, *Funaria* sp. – producers

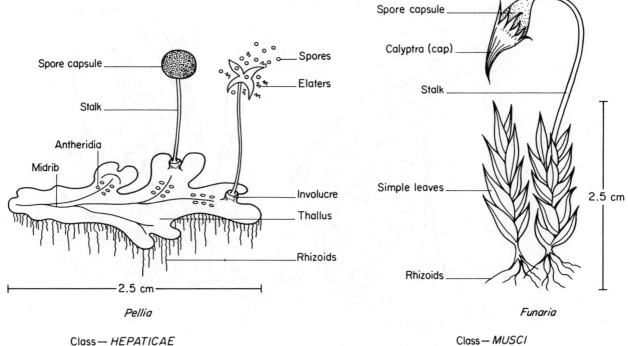

Pellia

Class— *HEPATICAE*

Funaria

Class— *MUSCI*

Figure 2.1 *(continued)*.

(e) Structure of a fern — producer

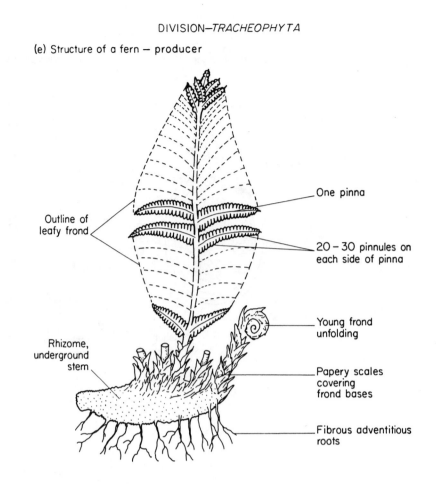

Outline of
leafy frond

One pinna

20 – 30 pinnules on
each side of pinna

Young frond
unfolding

Rhizome,
underground
stem

Papery scales
covering
frond bases

Fibrous adventitious
roots

Class— *PTERIDOPHYTA*

(f) Branches of a gymnosperm,
Pinus sylvestris — producer

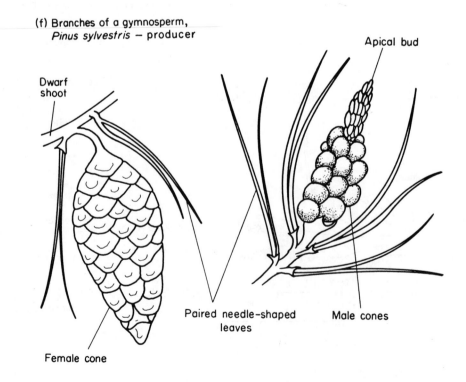

Apical bud

Dwarf
shoot

Paired needle-shaped
leaves

Male cones

Female cone

Class— *GYMNOSPERMAE*

Figure 2.1 *(continued).*

Subdivision — *SPERMATOPHYTA*

(g) Structure of a flowering plant — producer

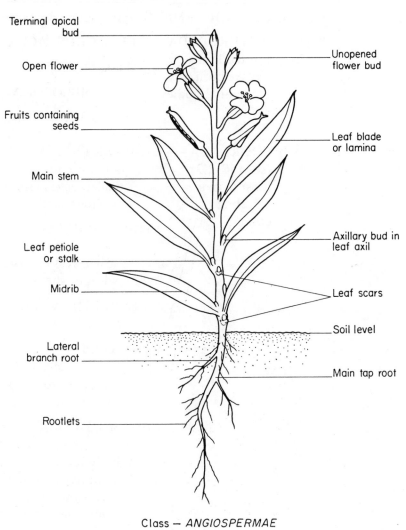

Terminal apical bud

Open flower

Fruits containing seeds

Main stem

Leaf petiole or stalk

Midrib

Lateral branch root

Rootlets

Unopened flower bud

Leaf blade or lamina

Axillary bud in leaf axil

Leaf scars

Soil level

Main tap root

Class — *ANGIOSPERMAE*

Figure 2.1 *(continued)*.

* 6. *Animal classification* can be summarised as follows:

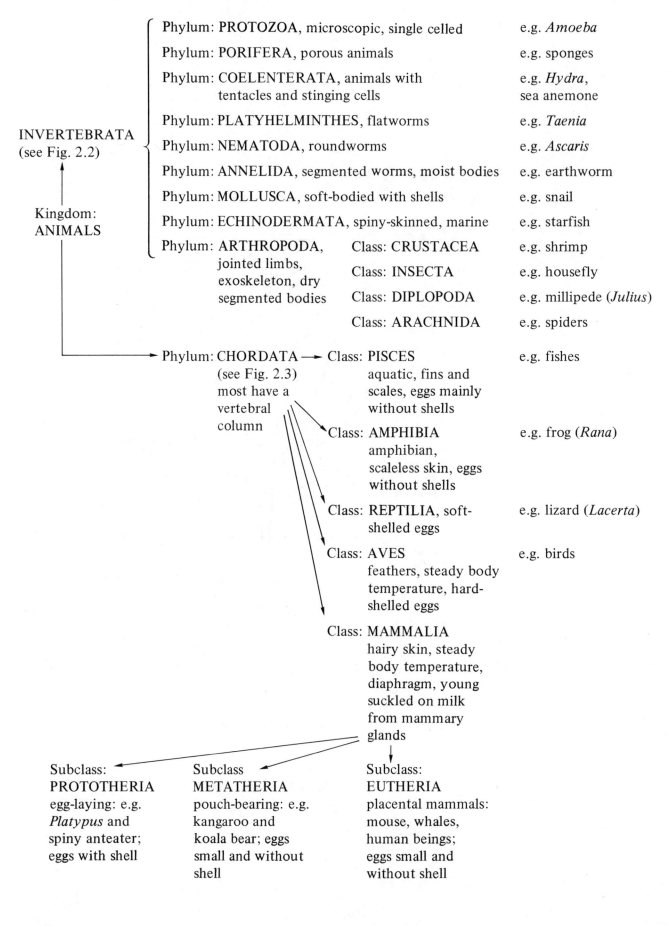

Kingdom:
ANIMALS

INVERTEBRATA
(see Fig. 2.2)

Phylum: PROTOZOA, microscopic, single celled	e.g. *Amoeba*
Phylum: PORIFERA, porous animals	e.g. sponges
Phylum: COELENTERATA, animals with tentacles and stinging cells	e.g. *Hydra*, sea anemone
Phylum: PLATYHELMINTHES, flatworms	e.g. *Taenia*
Phylum: NEMATODA, roundworms	e.g. *Ascaris*
Phylum: ANNELIDA, segmented worms, moist bodies	e.g. earthworm
Phylum: MOLLUSCA, soft-bodied with shells	e.g. snail
Phylum: ECHINODERMATA, spiny-skinned, marine	e.g. starfish

Phylum: ARTHROPODA, jointed limbs, exoskeleton, dry segmented bodies

Class: CRUSTACEA	e.g. shrimp
Class: INSECTA	e.g. housefly
Class: DIPLOPODA	e.g. millipede (*Julius*)
Class: ARACHNIDA	e.g. spiders

Phylum: CHORDATA (see Fig. 2.3) most have a vertebral column

Class: PISCES aquatic, fins and scales, eggs mainly without shells — e.g. fishes

Class: AMPHIBIA amphibian, scaleless skin, eggs without shells — e.g. frog (*Rana*)

Class: REPTILIA, soft-shelled eggs — e.g. lizard (*Lacerta*)

Class: AVES feathers, steady body temperature, hard-shelled eggs — e.g. birds

Class: MAMMALIA hairy skin, steady body temperature, diaphragm, young suckled on milk from mammary glands

Subclass:
PROTOTHERIA
egg-laying: e.g.
Platypus and
spiny anteater;
eggs with shell

Subclass
METATHERIA
pouch-bearing: e.g.
kangaroo and
koala bear; eggs
small and without
shell

Subclass:
EUTHERIA
placental mammals:
mouse, whales,
human beings;
eggs small and
without shell

PHYLUM *PROTOZOA*

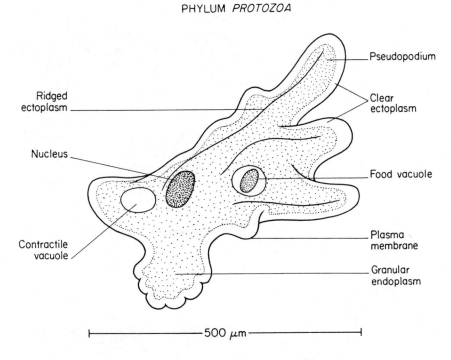

Pseudopodium

Ridged
ectoplasm

Clear
ectoplasm

Nucleus

Food vacuole

Contractile
vacuole

Plasma
membrane

Granular
endoplasm

—500 μm—

(a) Amoeba – consumer and detritivore

PHYLUM *PORIFERA (SPONGES)*

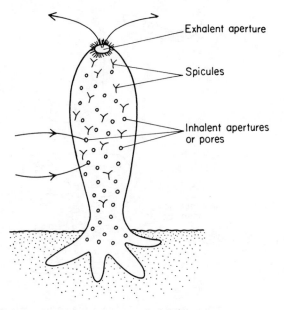

Exhalent aperture

Spicules

Inhalent apertures
or pores

(b) Sponge – consumer and detritivore

Figure 2.2   Examples of invertebrates.

PHYLUM *COELENTERATA*

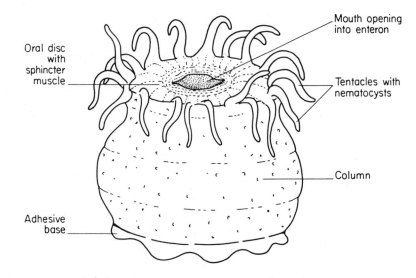

Oral disc with sphincter muscle

Mouth opening into enteron

Tentacles with nematocysts

Column

Adhesive base

(c) Sea anemone — consumer and omnivore

PHYLUM *PLATYHELMINTHES*

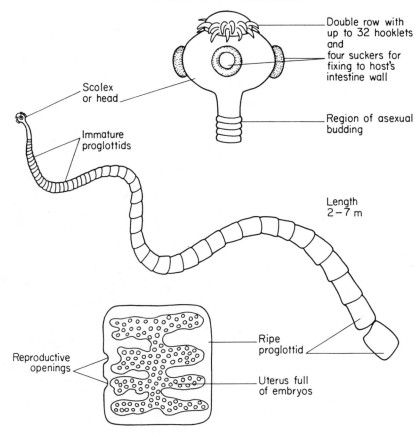

Double row with up to 32 hooklets and four suckers for fixing to host's intestine wall

Scolex or head

Region of asexual budding

Immature proglottids

Length 2 – 7 m

Reproductive openings

Ripe proglottid

Uterus full of embryos

(d) Tapeworm — consumer and parasite

Figure 2.2 *(continued).*

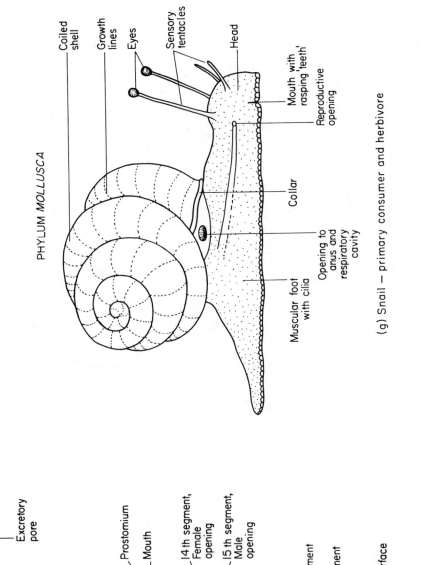

PHYLUM *MOLLUSCA*

Coiled shell

Growth lines

Eyes

Sensory tentacles

Head

Mouth with rasping 'teeth'

Reproductive opening

Collar

Opening to anus and respiratory cavity

Muscular foot with cilia

(g) Snail – primary consumer and herbivore

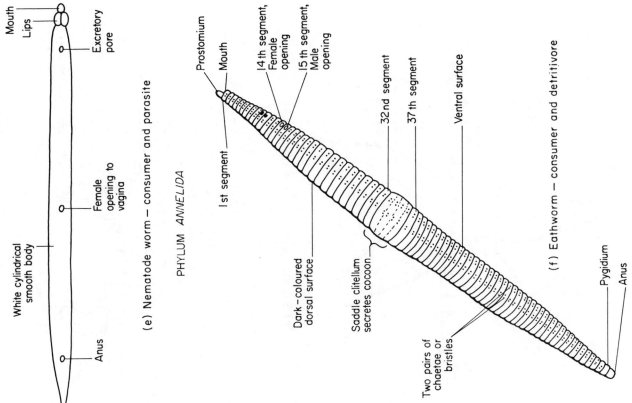

PHYLUM *NEMATODA*

Mouth

Lips

Excretory pore

White cylindrical smooth body

Female opening to vagina

Anus

(e) Nematode worm – consumer and parasite

PHYLUM *ANNELIDA*

Prostomium

Mouth

1st segment

14th segment, Female opening

15th segment, Male opening

Dark-coloured dorsal surface

Saddle clitellum secretes cocoon

Two pairs of chaetae or bristles

32nd segment

37th segment

Ventral surface

(f) Eathworm – consumer and detritivore

Pygidium

Anus

Figure 2.2 *(continued)*.

PHYLUM *ARTHROPODA*

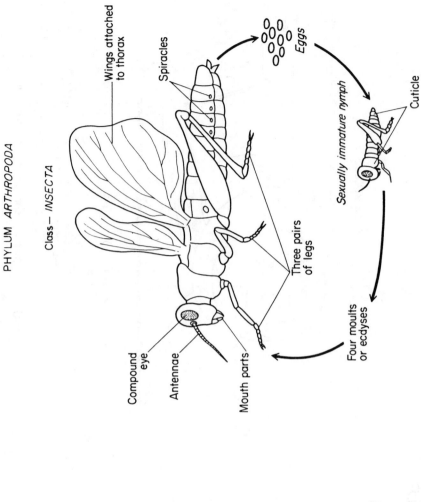

Class— *INSECTA*

Wings attached to thorax

Spiracles

Eggs

Sexually immature nymph

Cuticle

Three pairs of legs

Compound eye

Antennae

Mouth parts

Four moults or ecdyses

(j) Adult grasshopper – primary consumer and herbivore

Figure 2.2 *(continued).*

PHYLUM *ECHINODERMATA*

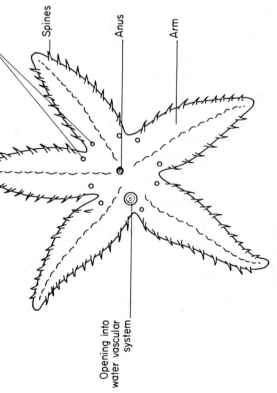

Genital openings

Spines

Anus

Arm

Opening into water vascular system

(h) Starfish – consumer and omnivore

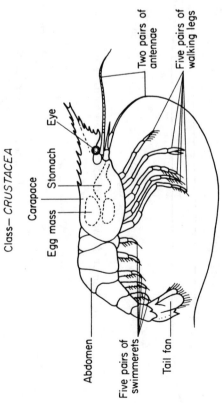

Class— *CRUSTACEA*

Two pairs of antennae

Five pairs of walking legs

Eye

Stomach

Carapace

Egg mass

Abdomen

Five pairs of swimmerets

Tail fan

(i) Shrimp (side view) – consumer and detritivore

Class—*DIPLOPODA*

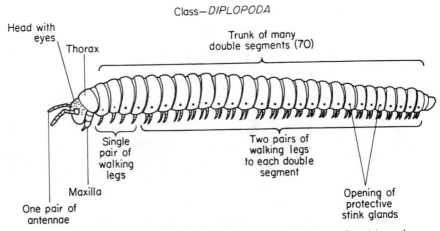

(k) Millipede (side view) — consumer and detritivore (herbivore)

Class—*ARACHNIDA*

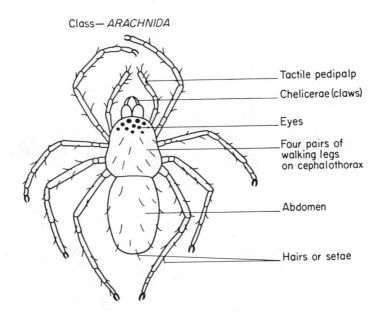

(l) Spider (dorsal view) — consumer and carnivore

Figure 2.2 *(continued).*

Class—PISCES

CARTILAGINOUS

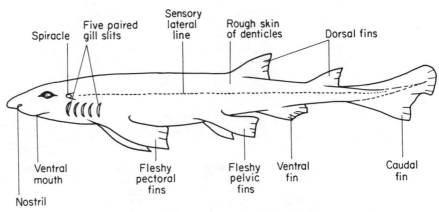

(a) Dogfish (60 cm) — consumer and carnivore

Figure 2.3 Examples of the phylum Chordata.

BONY

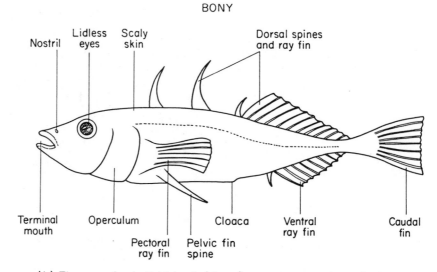

(b) Three – spined stickleback (6 cm) – consumer and carnivore

Class–AMPHIBIA

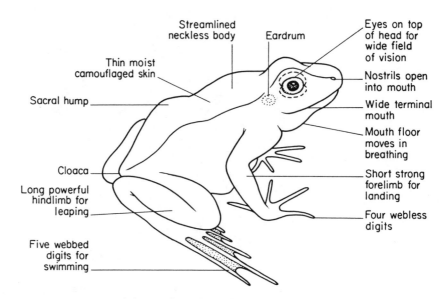

(c) Frog (8 cm) – consumer, omnivore

Class–REPTILIA

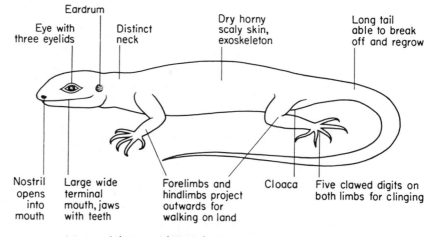

(d) Lizard (18 cm) – consumer, carnivore

Figure 2.3 *(continued).*

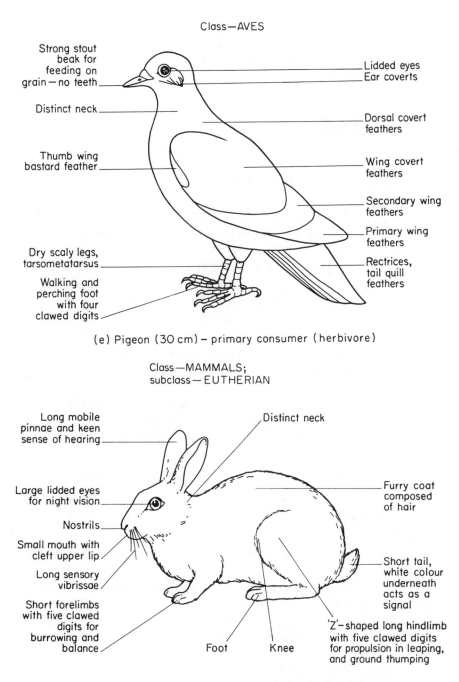

Class—AVES

Strong stout beak for feeding on grain—no teeth

Distinct neck

Thumb wing bastard feather

Dry scaly legs, tarsometatarsus

Walking and perching foot with four clawed digits

Lidded eyes
Ear coverts

Dorsal covert feathers

Wing covert feathers

Secondary wing feathers

Primary wing feathers

Rectrices, tail quill feathers

(e) Pigeon (30 cm) – primary consumer (herbivore)

Class—MAMMALS;
subclass—EUTHERIAN

Long mobile pinnae and keen sense of hearing

Large lidded eyes for night vision

Nostrils

Small mouth with cleft upper lip

Long sensory vibrissae

Short forelimbs with five clawed digits for burrowing and balance

Distinct neck

Furry coat composed of hair

Short tail, white colour underneath acts as a signal

Foot   Knee

'Z'-shaped long hindlimb with five clawed digits for propulsion in leaping, and ground thumping

(f) Rabbit (30 – 45 cm) – **primary consumer (herbivore)**

Figure 2.3 *(continued).*

(c) Body Terminology

1. *Radial symmetry* is the division of the body into similar halves when the body is cut across a *diameter*. It is seen in Coelenterata (such as the sea anemone) and Echinodermata (for example, sea urchins), which are drifting and sedentary (fixed) animals. Certain flowers are radially symmetrical.

2. *Bilateral symmetry* is the division of the body into similar halves when the body is cut along *one median plane*. Seen in free moving animals of all kinds. Bilaterally symmetrical bodies have the following surfaces or positions:

   Lateral, left or right lateral sides.

   Anterior or fore or front end, which leads in locomotion.

   Posterior or hind end, which follows or points backwards in locomotion.

Dorsal is the upper or upwards-directed surface in all animals, except in human beings when it is the backwards-directed surface.

Ventral is the lower or downwards-directed surface in all animals, except in human beings when it is the forwards-directed surface.

2.2 Practical Work

(a) Species Identification

1. *Keys* are used to identify an organism from its *external* features and to assign it to a *genus* and *species*. Once a plant is identified as being a member of a named *class*, e.g. fungi, mosses or angiosperms, or an animal is identified as being a member of a named *phylum* or *class*, e.g. Coelenterata or fishes, it can then be specifically identified with specially prepared keys.
2. *Specific identification* requires a knowledge of:
   (i) the phylum or class;
   (ii) terms used in naming structures and forms.
3. *Habitat species identification*. Special keys are constructed for identification of all organisms found in particular habitats, such as a sea shore, a freshwater pond, or woodland.
4. *Method of key use*. Most identification keys involve recognition of characteristic structures, number of parts, colours, position, etc. Chemical tests may be used, for example in keys for identifying lichens.
5. Drawing *diagrams* is important to indicate gross and fine details, or differences, present in a species specimen. (See page 283.)

Example

Sea fish identification key

Start at No. 1 on the left-hand side and follow the number or solution on the right-hand side. (See also key in objective questions 10 and 11.)

	Eel-like body, no paired fins, round sucking mouth	Lamprey, hagfish
1		*Go to key 1*
	Body not of that shape	2
	No gill cover, 5 or 6 gill slits	Shark, ray
2		*Go to key 2*
	Gill covers present	Bony fish
		Go to key 3

KEY 1: CYCLOSTOMES

This would be a key to identify lampreys and hagfish.

KEY 2: CARTILAGINOUS FISH

This key would lead to identification of 3000 different species of marine fish.

KEY 3: BONY FISH

This key includes up to 20 000 species of freshwater and marine fish.

Flora

These are keys for the identification of some of the 300 000 species of seed plants.

2.3 Examination Work

(a) Multiple-choice Objective Questions

1. Which of the following is a *correct* statement concerning Protozoa?
   (a) They are all parasites.
   (b) They are the simplest animals.
   (c) They are autotrophic.
   (d) They are all unicellular organisms.

2. Which of the following is true concerning a *Spirogyra* cell?
   (a) The cell wall is semi-permeable.
   (b) The cell wall is of cellulose.
   (c) The cell wall contains chlorophyll.
   (d) The cell wall is a lipid substance.

3. Which one of the following is a characteristic common to birds and mammals?
   (a) teeth
   (b) hair all over the body surface
   (c) constant body temperature
   (d) external ear pinnae

4. Which one of the following is an insect?
   (a) locust
   (b) shrimp
   (c) centipede
   (d) spider

5. Which of the following is a characteristic of all adult members of the sub-phylum Vertebrata?
   (a) bilaterally symmetrical bodies
   (b) scaly skins
   (c) fins for limbs
   (d) constant body temperatures

6. The question that follows refers to the following information, listing characteristics of four species of plants.

	I	II	III	IV
Normal habitat	land	land	land	aquatic
Vascular tissue present	yes	yes	no	no
Flowers produced	no	yes	no	no
Seeds produced	yes	yes	no	no
Autotrophic	yes	yes	no	yes

Which species is most likely to be a fungus?
(a) I
(b) II

(c) III
(d) IV

Questions 7, 8 and 9:
The following animals were collected from a habitat and had the following characteristics:
Animal A. hair, backbone, claws
Animal B. feathers, backbone, claws
Animal C. hair, backbone, no claws
Animal D. scales, backbone, claws
Animal E. shell, muscular foot, gills

7. The animals A to D inclusive belong to the group of animals called:
   (a) echinoderms
   (b) invertebrates
   (c) vertebrates
   (d) mammals

8. Animal E is most likely to belong to the
   (a) molluscs
   (b) annelids
   (c) echinoderms
   (d) crustaceans

9. The two animals most closely related and members of the same class are:
   (a) B and D
   (b) A and B
   (c) B and C
   (d) A and C

* Questions 10 and 11 are based on the following key:
  1. (a) Wings present 2
     (b) Wings absent . . . . .Order Apterygota
  2. (a) With one pair of wings. . . . . .Order Diptera
     (b) With two pairs of wings. 3
  3. (a) Front wings of coarser texture than hind wings. .4
     (b) All wings membranous. May be hair or scale covered 8
  4. (a) Basal two-thirds of front wing thickened, remainder
         membranous . . . . Order Hemiptera
     (b) Whole of front wing of same texture. 5
  5. (a) Front wings hard and horny . . . . Order Coleoptera
     (b) Front wings slightly thickened with distinct veins. 6
  6. (a) Mouthparts of piercing type . . . . Order Hemiptera
     (b) Mouthparts of biting type 7
  7. (a) Hind legs much longer than other legs. . . . . .Order Orthoptera
     (b) All legs more or less equal in length. . . . . .Order Blattoidea
  8. (a) Wings and body completely covered by fine
         scales or hairs . . . . Order Lepidoptera
     (b) Wings without scales or hairs. 9
  9. (a) Hind and front wings linked by a row of hooks. Front of
         abdomen narrowed to form a 'waist'. . . . . .Order Hymenoptera
     (b) Wings not joined. No 'waist' . . . . .Order Odonata

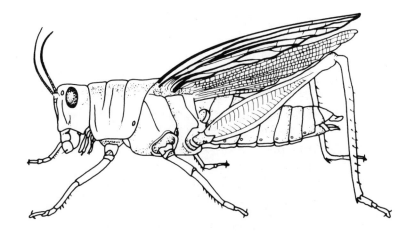

✳ 10. According to the key, which of the following orders does the insect shown in the diagram belong to?
    (a) Diptera
    (b) Odonata
    (c) Hemiptera
    (d) Orthoptera

(Western Australia Tertiary Admissions Examination 1983)

✳ 11. Lepidoptera would have the following characteristics:
    (a) two pairs of membranous wings covered in fine scales
    (b) one pair of membranous wings lacking scales or hairs
    (c) two pairs of wings, the front pair being of coarse texture
    (d) two pairs of membranous wings which are not hooked together and lack hairs or scales

(Western Australia Tertiary Admissions Examination 1983)

12. Two particular plants are classified into the same class. To which one of the following groups would the plants belong?
    (a) family
    (b) division
    (c) genus
    (d) species

13. Which two of the following British flowering plants listed by their botanical names would look most alike: I, *Agrostis gigantea*; II, *Ranunculus arvensis*; III, *Lamium album*; IV, *Agrostis canina*; V, *Myosotis arvensis*;
    (a) I and III
    (b) II and V
    (c) III and V
    (d) IV and I

14. Which one of the following is a member of the subdivision Bryophyta?
    (a) a pine tree
    (b) a moss
    (c) a fern
    (d) a buttercup

15. Which one of the following is *not* a member of the division of plants Tracheophyta?
    (a) a mushroom
    (b) a bracken plant
    (c) a larch tree
    (d) an onion plant

16. Radial symmetry refers to an animal:
    (a) without regular shape in any plane
    (b) which has separate body segments
    (c) that can be divided into similar halves in any diameter
    (d) with a backbone or vertebral column

17. If an animal has a ventral nerve cord it will be near its:
    (a) lower ventral surface
    (b) upper dorsal surface
    (c) head
    (d) tail

18. Which pair of characteristics do members of the phylum Coelenterata have?
    (a) body segmentation and exoskeleton
    (b) radial symmetry and exoskeleton
    (c) endoskeleton and bilateral symmetry
    (d) radial symmetry and stinging cells

19. A member of the phylum Annelida would:
    (a) be parasitic and non-moving
    (b) have an exoskeleton and bilateral symmetry
    (c) have body segmentation and no skeleton
    (d) have radial symmetry and an endoskeleton

20. Members of the phylum Arthropoda differ from other phyla by having:
    (a) an exoskeleton and body segmentation
    (b) a ventral nerve cord and endoskeleton
    (c) an endoskeleton and jointed appendages
    (d) an exoskeleton and radial symmetry

(b) Structured Questions

Question 2.1

Referring to the accompanying photograph:

(a) Indicate the classification of the animal in the photograph: (i) phylum; (ii) class. **(2 marks)**

(b) State two reasons why the animal is a member of the class in (ii). **(2 marks)**

(c) Name one *internal* structure that would show that this animal belongs to the phylum in (i) **(1 mark)**

(d) What does this animal feed upon when first born? **(1 mark)**

Question 2.2

The diagrams represent examples of animals from the phylum Arthropoda and the phylum Chordata.

(a) Separate the six animals shown into two groups by putting the appropriate letters (A, B, C, D, E or F) under the correct heading.

(b) Name each class to which animals A, C and F belong.

(c) Name one feature which is characteristic of all animals in the Arthropoda, but of no other group.

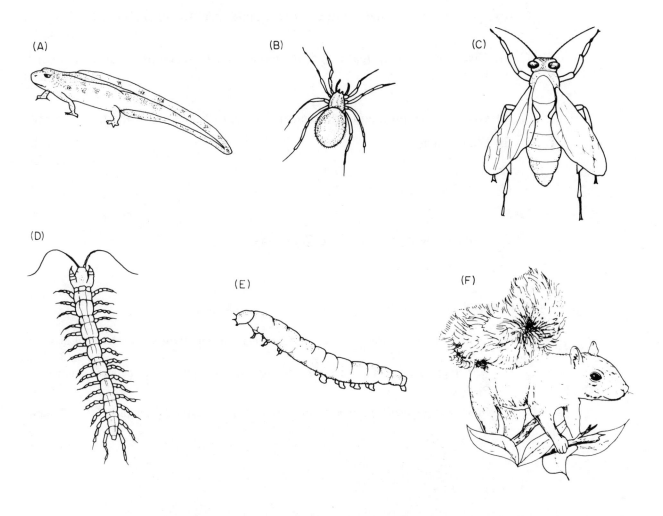

(A) (B) (C)

(D) (E) (F)

(d) Name one class in the Arthropoda which is *not* represented in the diagram.

(e) Name one feature possessed by E which is *not* present later in its life cycle. **(4 marks)**

Question 2.3

Mucor is a fungus, the main body of which consists of a number of branching 'threads'. It is often found growing on damp bread.

(a) (i) What is the name given to an individual thread? (ii) What is the collective name of all the 'threads'?

(b) Name the type of nutrition displayed by *Mucor*. **(3 marks)**

(AEB, 1981)

(c) Free-response-type Questions

(i) *Short-answer Free-response Questions*

1. By what features would you recognise an insect? **(6 marks)**

2. Give the name of the microscopic living threads that make up the body of a fungus. **(1 mark)**

3. (a) Name a species of insect. **(1 mark)**

   (b) To which phylum does it belong? **(1 mark)**

   (c) Give three external features that enable you to recognise it as an insect. **(3 marks)**

   (d) Give two external features that enable you to recognise it as a member of its phylum. **(2 marks)**
   (OLE)

4. (a) Give the common and full scientific name of a species of plant. **(2 marks)**

   (b) What is a species? **(2 marks)**
   (OLE)

(ii) *Long-answer Free-response Questions*

✳ Question 2.4

(a) Describe the structure of a cell of *Spirogyra* and an *Amoeba* with labelled diagrams. **(8 marks)**

(b) With these organisms as examples, discuss the basic differences between animals and plants. **(5 marks)**

(c) In what ways is *Mucor* similar to and different from a typical green plant? **(2 marks)**
(SUJB)

Answer to Q2.4

This is a *direct* question, with all the information being found in one chapter or section of a textbook.

(a) Two large, well labelled, simply drawn *diagrams* are needed in this first part. There is no need for a written description. This is a question in which you should be able to draw and label the diagram correctly from memory. The diagrams in Figs 2.1, 2.2 and 2.3 are all that are needed.

(b) The key term here is 'differences between plants and animals' found in 2.1(b)4, but answered with respect to *Spirogyra* and *Amoeba*. A *table of comparison* is the best way to set out this answer neatly and quickly (see page 40).

(c) Since *Mucor* is a fungus, a comparison is expected with *all* other green or chlorophyll-containing plants. Give ONE example in each case, since the question uses the words 'similar' and 'different' (see page 40).

Answer to Q2.4(b)

Plant: Spirogyra	Animal: Amoeba
1. Green chloroplasts	1. No chloroplasts
2. Autotrophic: photosynthesis of organic food	2. Heterotrophic nutrition with ingestion of ready-made organic food
3. No excretory system	3. Excretory vacuole vital to eliminate toxic metabolic waste and regulate osmosis
4. No locomotion	4. Slow locomotion of whole body
5. No irritability	5. Irritability and responses shown to various stimuli
6. Cellulose wall	6. Ectoplasm – lipid/protein membrane only, no cell wall
7. Large vacuole	7. Numerous small vacuoles

Answer to Q2.4(c)

	Mucor: fungus	Green plants
Difference	No chlorophyll (heterotrophic)	Chlorophyll present (autotrophic)
Similarity	Cellular with cell walls	Cellular with cell walls

✶ **Question 2.5**

(a) Name a filamentous or colonial alga. **(1 mark)**

Name a mould. **(1 mark)**

Give two structural features which the alga you have named does not share with the mould you have named. **(2 marks)**

(b) Name an example of each of the following types of organism. For each example you have named, give two features not shared with either of the other two: (i) a unicellular animal; (ii) a unicellular plant; (iii) a bacterium. **(9 marks)**

(c) In the system of classification of organisms, the following terms are used: class, family, genus, kingdom, order, phylum, species. Complete the following list by inserting the missing terms in correct order:
---------- Phylum ---------- Order ---------Genus **(4 marks)**
(OLE)

Answer to Q2.5

This is a *direct* and partly a structured question answerable with information from one source, a chapter or class notes.

(a) Filamentous alga *Spirogyra* (colonial alga *Volvox*). Named mould *Mucor* or *Pythium*.
Structural features of alga not shared with mould or fungus:
(i) chloroplast and chlorophyll present
(ii) cell wall cellulose (in fungi, cell wall is mostly hemicellulose, chitin, or lipid and protein)

(b)

	Unicellular animal: Amoeba	Unicellular plant: Chlamydomonas	Bacterium: Bacillus
(i)	Cell membrane lipid/protein	Cellulose wall	Cell wall mainly protein and/or polysaccharide
(ii)	Several small vacuoles for food and excretion	Two contractile vacuoles	No vacuoles

(c) *Seven* terms are listed in the structured question; the missing four are arranged in the answer in the following order: *kingdom* – phylum – *class* – order – *family* – genus – *species*.

∗ Question 2.6

(a) Compare the features that distinguish living organisms from non-living objects. **(9 marks)**

(b) Draw a labelled diagram to show the structure of a *named* annelid animal and a *named* algal plant. **(9 marks)**

(c) List the main differences between the two named organisms in (b). **(2 marks)**

Answer to Q2.6

A *direct* question.

(a) This part requires a table to compare the characteristic features of living and non-living things, and you need to name examples.

Living organisms (e.g. mouse)	Non-living objects (e.g. stone)
Movement	No movement (except vehicles)
Feeding	No feeding
Growth	No growth (except crystals)
Respiration	No respiration
Reproduction	No reproduction
Response	No responsiveness (except photo-electric cells)
Excretion	No excretory process

(b) Two large labelled diagrams are needed here, both drawn correctly from memory. The annelid animal is earthworm (Fig. 2.3(f)), and the algal plant *Chlamydomonas* or *Spirogyra*, shown in Fig. 2.1(c)

(c) This requires a comparison of differences between the annelid worm and the alga. Two differences are sufficient. Many more differences could be given, including cellulose wall, locomotion, multicellular, segmentation, etc.

Annelid worm	Alga
No green chloroplasts	Green chloroplasts present
Heterotrophic nutrition	Autotrophic nutrition

2.4 Self-test Answers to Objective and Structured Questions

Answers to Multiple-choice Objective Questions

1. d 2. b 3. c 4. a 5. a 6. c (I – conifer, II – angiosperm, IV – algae)
7. c 8. a 9. d 10. d (this is an example of a key to identify orders of insects)
11. a 12. b 13. d 14. b 15. a 16. c 17. a 18. d 19. c 20. a

Answer to Structured Question 2.1

(a) (i) phylum Chordata (vertebrate is *not* the correct name)
     (ii) class Mammalia or mammals

(b) (i) hair
     (ii) external ears – pinnae

(c) Internal structure typical of phylum Chordata – vertebral column.

(d) Milk from mammary glands.

Answer to Structured Question 2.2

This is an example of a question based on classification.
(a) Arthropoda = B, C, D, E
    Chordata = A, F

(b) Class A = Amphibia
    Class C = Insecta
    Class F = mammals

(c) Characteristic feature of all Arthropoda – jointed paired legs and hard exo-skeleton.

(d) Arthropoda not represented are the crustaceans, e.g. lobsters and crabs.

(e) This is a larva which is without tube feet or prolegs in its later life cycle (see Growth, Chapter 8).

Answer to Structured Question 2.3

(a) (i) hypha (plural – hyphae)
     (ii) mycelium

(b) Since *Mucor* feeds on organic food (bread), nutrition is *heterotrophic* and *saprophytic*.

Theme II

Structure and Functioning of Living Organisms

3 Cells, Tissues, Organs and Organisation

3.1 Theoretical Work Summary

(a) Cells

1. *Unicellular* organisms consist of *one* cell, and include protozoa, bacteria, blue-green algae and certain other algae.
2. *Acellular* or non-cellular organisms or *tissues* not composed of cells:
    (a) Plant *coenocyte*, cell substance with many nuclei surrounded by one cell wall — found in certain fungi, e.g. *Mucor*, and certain algae, e.g. *Vaucheria*.
    (b) Animal *syncytium*, cell substance with many nuclei surrounded by one cell membrane — found in certain tissues, e.g. striated skeletal and heart muscle.
3. *Multicellular* organisms are made up of *many* cells.

(b) Cell Structure

1. The *generalised structure* of a typical higher animal and higher plant cell or *eukaryote* cell is shown in Fig. 3.1.
2. *Eukaryotes* include algae, fungi, protozoa and multicellular higher plants and animals.
3. *Prokaryotes* include bacteria and blue-green algae only.
4. *Protoplasm* is the name given to the living cell's contents:
    (a) plasma or cell *membrane* composed of lipids and proteins
    (b) *cytoplasm* containing cell sap
    (c) *nucleus*
    (d) vacuole or vacuoles
5. *Cell wall*: non-living, rigid material surrounding the cell protoplasm of bacteria and plants. (*No* cell walls in ANIMAL cells.)
6. *Biological membranes* are of three kinds:
    (a) *Cell membrane*: plasma membrane or plasmalemma, or the cell surrounding membrane.
    (b) *Nuclear membrane* surrounding the cell nucleus.
    (c) *Organelle membranes* surrounding or forming cell organelles with specific functions.
7. *Organelle*: specialised parts of living cells. Examples are the nucleus, mitochondria, ribosomes, Golgi bodies and chloroplasts.
8. The differences between the cells of simple organisms (prokaryotes) and all other organisms (eukaryotes) are summarised in Table 3.1.

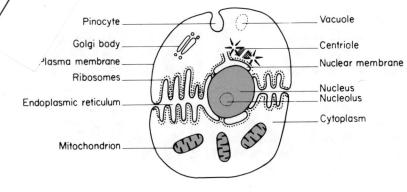

(a) Generalised animal cell

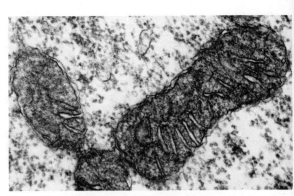

Figure 3.2 Electron micrograph of the mitochondria present in the egg cell of *Pomatoceros* sp., a small marine worm living within tubes fixed to rock surfaces on the sea-shore. The mitochondria are magnified 45 000 times; the larger organelle measures about 7 μm by 2 μm. Reproduced by courtesy of Iolo Ap Gwynn, University College of Wales, Aberystwyth.

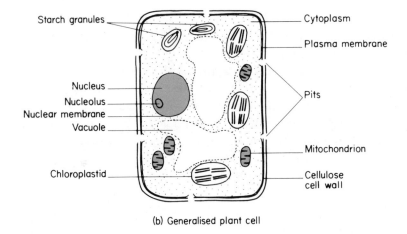

(b) Generalised plant cell

Figure 3.1 Generalised structure of (a) an animal cell and (b) a plant cell, to show the organelles. In examinations you will probably be asked to 'label and explain these figures'.

Table 3.1 A comparison of prokaryotes and eukaryotes

Feature	Simple organisms: bacteria, blue-green algae and viruses (prokaryotes)	All other organisms (eukaryotes)
Cell size	Up to 3 μm diameter	Up to 40 μm diameter
Cell wall	Amino acids and polysaccharides. (No cell wall in viruses.)	Cellulose only in green plants. Fungal cellulose in fungi
Nucleus	No true nucleus; no nucleolus; no chromosomes; no centriole	True nucleus present with chromosomes and nucleolus; centriole in higher animals only
Organelles	Few, without a membrane	Many, with a double or single membrane. No chloroplasts in animals

9. Table 3.2 compares the differences between higher (eukaryote) plant ⸝
animal cells.

Table 3.2 A comparison of eukaryote plant and animal cells

Feature	Higher plant cells	Higher animal cells
Cell wall	Rigid cellulose wall present — may also be *lignified*	None
Vacuole	Large and permanent, forming up to 80% of cell volume	Small, numerous and temporary if present
Centrioles	In algae but *not* in higher plants	Present
Chloroplasts	Present in chlorenchyma cells	Absent
Granules	Starch	Glycogen
Cilia	Absent	Present in certain cells

(c) Tissues and Organs

1. *Tissues* are groups of cells in multicellular organisms specialised to perform a specific or *specialised* function.
2. *Organs* are composed of more than one tissue, forming a structural and functional unit in a multicellular organism. Examples of *plant* organs are the root, stem, leaf, bud and flower; *animal* organs include the heart, stomach, kidney, liver and skin.
3. *Organ systems* are composed of different organs working together towards one main function of maintaining life in the organism.

    Flowering plants have two organ systems:
    (a) *vegetative*: root, stem, leaf and bud, for nutrition and growth
    (b) *reproductive*: flowers, fruit and seeds

    Mammals have nine main organ systems:
    (a) *respiratory*: larynx, trachea, lungs
    (b) *alimentary*: teeth, tongue, gastrointestinal tract
    (c) *urinary*: kidney and bladder
    (d) *reproductive*: male and female reproductive organs
    (e) *nervous*: brain, nerves, sensory organs
    (f) *muscular*: skeletal muscles
    (g) *circulatory*: heart, arteries, veins and lymph
    (h) *skeletal*: bones, joints
    (i) *homeostatic*: kidney, liver, pancreas, skin

4. Table 3.3 summarises the tissues to be found in vascular plants (Tracheophyta); see also Fig. 3.3.

(d) Organisation

1. *Unicellular* organisms perform the different functions essential to maintaining life in various organelles within the cytoplasm of a single cell.

    ATOMS⟶MOLECULES⟶ORGANELLES⟶UNICELLULAR ORGANISM

51

* Table 3.3 Structure and function of tissues in vascular plants

Tissues and location	Structure	Function
1. *Apical meristem*: primary in stem and root tip (apex)	Small, living, actively dividing cells, thin walls, no vacuoles, large nucleus	Primary growth and development
2. *Epidermal*: outer layer of root, stem and leaf *cork* in thickened woody stems and roots	Living cells, with slightly thickened cellulose walls, and waxy outer *cuticle*. Cork is thickened with *suberin*	Covering, protection and waterproofing
3. *Parenchyma*: a packing tissue between other tissues in cortex and pith	Living cells with thin walls, and with starch granules and numerous intercellular air spaces	Non-specialised and varied, e.g. storage and transport
4. *Chlorenchyma*: photosynthetic tissue mainly in leaves and green stems	Living cells, with thin walls and numerous chloroplasts, and intercellular air spaces	Photosynthesis, gas exchange
5. *Supporting tissues*: (i) *Collenchyma*, usually beneath epidermis	Living, elongated cells with cellulose thickening in cell corners	Mechanical support in herbaceous plants
(ii) *Sclerenchyma* found close to vascular tissue	*Non-living*, elongated cells of great length with cellulose and *lignin* thickening	Mechanical support as *fibres*
6. *Vascular tissue*: (i) *Cambium*, a lateral meristem inside the vascular bundle	Similar to apical meristem	Secondary growth in perennials
(ii) *Xylem* in the vascular bundle	Xylem vessels and tracheids: *non-living*, elongated cells, heavily *lignified* and of considerable length	Water and ion transport; support
(iii) *Phloem* in the vascular bundle	Sieve tubes and companion cells, thin unthickened walls, living, elongated cells	Transport of products of photosynthesis
7. *Gland tissue*: surface or internal	One or more cells grouped together, secreting a variety of products: nectar, essential scent oils, water, resin, rubber latex, gums and crystals	Variable function in pollination, water balance and excretory products

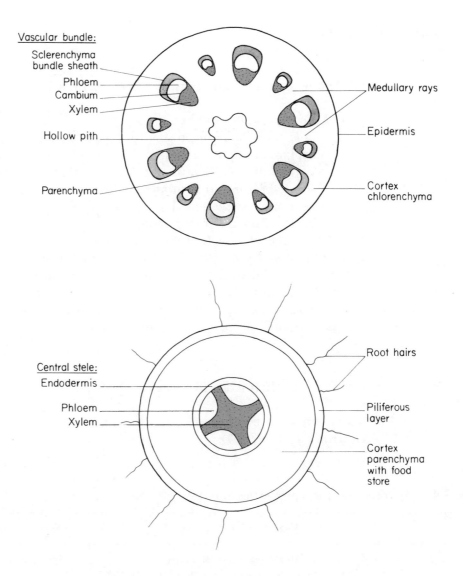

Figure 3.3   The arrangement of tissues in (top) the stem, and (bottom) the root of a butter-cup, *Ranunculus* sp., as seen in transverse section. In examinations you will probably be asked to 'label and explain these figures'.

2. *Simple multicellular* organisms, e.g. *Hydra* and the Thallophyta, group the body cells with their functional organelles into simple *tissues* performing one or more functions, which may be grouped with other tissues to form an *organ*, e.g. a tentacle with stinging cells, or a reproductive organ.

ATOMS⟶MOLECULES⟶ORGANELLES⟶CELLS⟶TISSUE ⟶ORGAN

SIMPLE MULTICELLULAR ORGANISM ⟵

3. *Higher multicellular* organisms, e.g. flowering plants and mammals, extend the level of organisation from tissues and simple organs into *organ systems* made up of several organs.

ATOMS ⟶MOLECULES ⟶ ORGANELLES ⟶ CELLS ⟶TISSUES

HIGHER MULTICELLULAR ORGANISM ⟵ ORGAN SYSTEM⟵ ORGANS

3.2 Practical Work

(a) Microscopical Preparations (See Section 15.5(d))

1. *Biological specimens* can be:
    - (a) *whole* unicellular organisms, or strips of tissue one cell thick — cheek cell scrapings or onion leaf scale:
    - (b) *sectional multicellular* material cut as thin slices or sections with a sharp flat-sided razor or *microtome* machine, in one of three planes:
        - (i) transverse section, TS
        - (ii) longitudinal or radial longitudinal section, LS or RLS
        - (iii) tangential section across a radius
2. *Hardening and preserving* involves treating the section with alcohol (ethanol).
3. *Staining* is needed since most biological material is transparent or translucent, with little contrast between tissues and cell structure appearance.
4. *Mounting* is placing a specimen on a microscope slide in a jelly, gum, Canada balsam resin or propanetriol (glycerol) and then covering with a glass *cover slip.*

(b) Types of Microscope (See Section 15.5(c))

Microscopes can be grouped into three main categories: light, phase contrast and electron (see Table 3.4).

Table 3.4   Microscopes: their properties and uses

Type	Magnification and image	Specimen treatment
Light microscope	Magnification = Objective lens × Eyepiece lens. From 40 to 1500 times. Image coloured	Living or non-living; chemically stained
Phase contrast microscope	Similar magnification to light microscope. Contrasts or shows small visual differences in transparent unstained material	Living cells. No staining. Used to observe nucleus division and movement of flagella
Electron microscope, two types:	Magnification up to 250 000 times. Black and white image on TV screen, or electron micrograph	Non-living material. Coated with osmium, lead or uranium metal to stain specimen. Gold and palladium used for scanning treatment
(i) transmission	electrons pass *through* specimen	
(ii) scanning	electrons *reflected* on specimen surface	

(c) Using a Light Microscope

1. First examine object with *low-power* objective lens.
2. *Raise* the objective, before swinging the microscope nose-piece to high-power objective.
3. *Focus* with high power by racking *upwards* from a close position to slide.
4. *Concave* side of mirror if there is *no* substage condenser. *Plane* side of mirror if there *is* a substage condenser.

3.3 Examination Work

(a) Multiple-choice Objective Questions

1. Which of the following structures is present in plant cells but absent from animal cells?
   (a) cell membrane
   (b) cell wall
   (c) nucleus
   (d) cytoplasm

2. The light microscope can be used to observe:
   (a) protein molecules
   (b) water molecules
   (c) chromosomes
   (d) phosphate ions

3. Which one of the following is an incorrect statement concerning the *cell membrane*?
   (a) It prevents certain substances leaving a cell.
   (b) It surrounds all cells.
   (c) It allows substances to enter and leave the cell.
   (d) It holds cells together.

4. The formation of proteins from amino acids is carried out at the:
   (a) ribosomes
   (b) chloroplasts
   (c) mitochondria
   (d) nuclear membrane

5. Which one of the following is a correct statement concerning *tissues*?
   (a) They hold organs together.
   (b) They form the cell membrane.
   (c) They are composed of cells of one kind.
   (d) They are composed of cells with several functions.

6. Which one of the following is a correct statement concerning *organs*?
   (a) They are composed of cells with the same function.
   (b) They are not found in plants.
   (c) Unicellular animals have a few organs.
   (d) They perform a certain overall function.

7. The correct sequence, from the smallest to the largest units, is:
   (a) cells, organelles, tissues, systems, organs
   (b) cells, tissues, organs, organelles, systems
   (c) organelles, cells, tissues, organs, systems
   (d) organelles, organs, tissues, cells, systems

8. Single-celled organisms show great variety and yet they have many similarities because most of them are:
   (a) microscopic
   (b) parasitic
   (c) plants
   (d) animals

9. Which of the following microscope objective lenses will provide a magnification of 200 times with the 5 times eyepiece lens?
    (a) × 4
    (b) × 10
    (c) × 20
    (d) × 40

10. Which of the following eyepiece and objective lens combinations will allow you to see the greatest number of blood cells on a slide at one time?
    (a) objective × 40, eyepiece × 10
    (b) objective × 10, eyepiece × 10
    (c) objective × 4, eyepiece × 5
    (d) objective × 40, eyepiece × 5

11. The *centriole* is an organelle which is important in:
    (a) enzyme storage
    (b) animal cell nucleus division
    (c) protein synthesis
    (d) photosynthetic plant cells

12. Some of the characteristics of several types of cells are recorded in the following table. An asterisk (*) shows the structure present in each cell. Which cell is most likely to be a plant cell?

	Cell wall	Cilia	Vacuole	Chloroplast	Centrioles
Cell A		*			
Cell B				*	
Cell C		*	*		*
Cell D	*		*	*	

13. A picture of cells taken with the aid of an electron microscope is called a:
    (a) photograph
    (b) photomicrograph
    (c) micrograph
    (d) electron micrograph

14. The cell membrane of a typical higher plant cell lacks one of the following:
    (a) pores
    (b) flagella
    (c) cellulose
    (d) protein

15. Which one of the following is an animal tissue?
    (a) parenchyma tissue
    (b) chlorenchyma tissue
    (c) meristem tissue
    (d) connective tissue

16. One of the following is *not* a component of protoplasm:
    (a) cytoplasm
    (b) vacuole
    (c) cell nucleus
    (d) cell wall

17. Which of the following plant cells is lacking a nucleus?
    (a) parenchyma
    (b) companion cells
    (c) xylem vessels
    (d) chlorenchyma

18. Which of the following plant tissues can contain many nuclei in cytoplasm bounded by a single cell wall?
    (a) *Spirogyra* filament
    (b) *Mucor* mycelium
    (c) buttercup root cortex
    (d) onion bulb scale

19. Which one of the following kinds of tissues is found only in animals?
    (a) epidermal tissue
    (b) storage tissue
    (c) contractile tissue
    (d) secretory tissue

20. Which of the following is near the maximum magnification by the electron microscope?
    (a) 3000 times
    (b) 30 000 times
    (c) 300 000 times
    (d) 3 million times

(b) Structured Questions

Question 3.1

The accompanying diagram shows a highly magnified yeast cell. Yeast is a fungus and has both plant-like and animal-like features.

Eight parts are labelled in the diagram. In a table as below write the name of each of the eight parts in the column you think most appropriate.

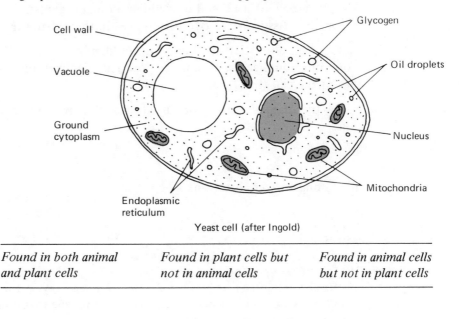

Yeast cell (after Ingold)

Found in both animal and plant cells	Found in plant cells but not in animal cells	Found in animal cells but not in plant cells

(8 marks)

(AEB, 1981)

Question 3.2

The diagram represents a transverse section of a young root from a flowering plant.

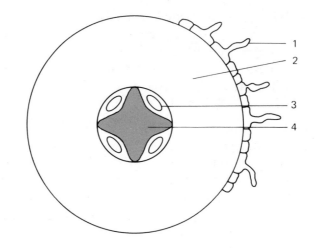

(i) State the function of the tissue numbered 1 in the diagram.

(ii) State *one* similarity and *one* difference in the function of the tissues numbered 3 and 4.

(iii) Give one further function of the root in addition to those mentioned in your answers to parts (i) and (ii). **(4 marks)**

(SUJB)

(c) Free-response-type Questions

(i) Short-answer Questions

1. Give four differences between animal cells and plant cells. **(4 marks)**

(L)

2. Give examples of a named unicellular animal organism and a cell from a named mammalian tissue. State two differences in their structure.

3. Name one cell structure that you could *not* see with the school light microscope but could see with a more powerful electron microscope. State the function of the cell structure.

4. Name the chemical substance used as a stain to show starch granules in plant cells.

(ii) Long-answer Questions

✳ Question 3.3

(i) Describe a typical plant or animal cell.

(ii) Name three differences between a typical plant and a typical animal cell.

(iii) Draw one specialised tissue cell and relate its structure to its function.

(iv) Why are cell membranes important? **(10 marks)**

(NISEC)

Answer to Q3.3

This is a *direct* question based on higher plant and animal (eukaryote) cells.

   (i) A labelled *diagram* (similar to Fig. 3.1) and brief written description are part of the question.
   (ii) Differences listed as follows:

Plant cell	Animal cell
Have a cell wall	Cell wall absent
Have chloroplasts	Chloroplasts absent
Large vacuole present	Small vacuoles may be present

Other differences: centrioles, cilia and starch granules.

   (iii) A diagram of *any* plant tissue selected from Table 3.3, or palisade cells or stomata cells from leaf structure described in Chapter 4, *plus* a note of structure as it is related to function. Select briefly from one of the following:

Function	Plant	Animal
Protection	Epidermis	Bone
Mechanical support	Collenchyma, sclerenchyma, xylem fibres and vessels	Bone Cartilage Connective tissue proper
Transport	Xylem vessels, sieve tubes	—
Contraction movement	—	Muscle
Impulse conduction	—	Neurone

   (iv) Cell membranes important as *one* of the following:
        (a) living *barrier*
        (b) *selectively permeable membranes* in allowing certain substances in and out of cell

∗ Question 3.4

   (a) Describe fully how you would prepare cells from the lining of your cheek for microscopic examination. **(4 marks)**

   (b) Explain how you would:
        (i) adjust the focus of a microscope; **(2 marks)**

        (ii) adjust the intensity of light passing through the object; **(2 marks)**
        (iii) increase the magnifying power of the microscope. **(3 marks)**

   (c) Make a labelled diagram to show the structure of a mammalian cell such as from the lining of your cheek. **(5 marks)**

   (d) Briefly describe the main stages in meiosis. **(5 marks)**

(e) Give one type of cell found in a mammal and one type of cell from another organism in which the nucleus is absent or greatly reduced. **(2 marks)**

(f) Give one type of cell found in a mammal and one type of cell from another organism in which the cytoplasm is absent or greatly reduced. **(2 marks)**
(OLE)

Answer to Q3.4

But for part (d) this would be a *direct* question. Part (d) refers to meiosis, which is normally dealt with in detail in sections relating to reproduction and genetics — Chapters 12 and 13.

(a) This involves microscopical preparation of human cheek *epithelial* cells
   (i) Collect first with a clean sterile spoon handle gently rubbed inside cheek.
   (ii) Place specimen on centre of clean microscope slide, and allow to dry.
   (iii) Add *methylene blue stain* to show up cell nuclei, removing excess by rinsing in clean water.
   (iv) Mount with a drop of propanetriol (glycerol) and cover with a cover slip.

(b) Microscope-use technique:
   (i) *Focus* by raising objective lens from a point close to cover slip.
   (ii) Light intensity is controlled by either: the *condenser iris diaphragm* or, if not fitted, by the *mirror*.
   (iii) Magnifying power increased from low to high power by exchanging objective lens say from × 4 to × 10 or × 20. This is done by either rotating the *nose-piece*, if fitted, or unscrewing the objective lens.

(c) The labelled diagram shown below includes the tiny mitochondria which are *just* visible with a high-power lens. Note the nucleolus clearly visible in the nucleus.

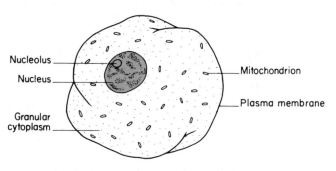

In examinations you will probably be asked to 'label and explain this figure'.

(d) *Meiosis* is two successive divisions of the nucleus to produce *four* daughter nuclei from *one* parent cell, each having *half* the number of chromosomes as the parent cell nucleus. (See section 13.1(c) and Fig. 13.1.)

(e) An example of a mammal cell without a nucleus is a red blood cell. Bacteria and blue-green algae (prokaryotes) have no true nucleus.

(f) Cytoplasm is greatly reduced in *mammal* connective tissue such as adipose tissue, cartilage or bone, also in *xylem vessels* and *sclerenchyma* fibres, in which cytoplasm is reduced or non-existent in 'dead cells'.

    (a) Describe by means of a fully labelled diagram the structure of a generalised higher plant cell as seen under the high power of a light microscope.

    (b) List the differences between a higher (eukaryote) plant cell and a generalised higher animal cell.

    (c) Explain how the differences between plant and animal cells is related to differences in structure and activity of complete higher plants and animals.

    (d) Describe how the different levels of functional organisation are built up, starting from atoms of chemical elements, in a multicellular higher plant organism.

Answer to Q3.5

(a) The generalised labelled diagram (Fig. 3.1) of a plant or animal cell is clearly an important diagram in biology.

(b) Cell differences are as follows:

Cell structure	Plant cell	Animal cell
Cell wall	Present	Absent
Chloroplasts	Present	Absent
Vacuoles	Single, large, central	Small, scattered
Granules	Starch	Glycogen
Centrioles	Absent	Present
Microvilli	Absent	Present

(c) Differences in structure and activity of complete higher animals and plants are as follows, the main differences being due to cell wall, method of nutrition (chloroplasts) and energy needs (mitochondria).

	Plants	Animals
Body structure	Spreading or branched with large surface area	Compact body with less surface area
Activity	Rooted or sessile, no need to search for food	Active and motile, essential to search for food
Reason	(i) Cellulose wall restricts movement and is *non-contractile* (ii) Food manufactured in chloroplasts from inorganic substances (iii) Fewer mitochondria due to *lower* energy needs	(i) Plasma membrane is flexible and contractile (ii) Food must be ready-made organic substances (iii) Many mitochondria due to *high* energy needs

(d) Briefly the following is required:

Element atoms ⟶ Molecules ⟶ Organelles ⟶ Cell ⟶ Tissues ⟶ Organ

⟶ Organ system   (vegetative and reproductive systems)   ⟶ multicellular higher plant organism

3.4 Self-test Answers to Objective and Structured Questions

Answers to Multiple-choice Objective Questions

1. b   2. c   3. c   4. a   5. c   6. d   7. c   8. a   9. d   10. c   11. b   12. d
13. d   14. c   15. d   16. d   17. c   18. b   19. c   20. c

Answer to Structured Question 3.1

Found in both animal and plant cells	Found in plant cells but not in animal cells	Found in animal cells but not in plant cells
Mitochondria	Cell wall	Glycogen
Nucleus	*Small lipid oil droplets	*Large lipid oil droplets
Endoplasmic reticulum	Large vacuole	
Ground cytoplasm		

Answer to Structured Question 3.2

(i)   Part numbered 1 is a *root hair* for absorption of soil water and ions.

(ii) Part 4 is the xylem tissue. Part 3 is the phloem tissue:

	Xylem tissue	Phloem tissue
Similarity	Transport	Transport
Difference	Water and ions *upwards*	Organic manufactured food *downwards*

(iii) Anchors plant in the soil. (Also food storage.)

4 Plant Nutrition

4.1 Theoretical Work Summary

(a) Autotrophic Nutrition

Autotrophic nutrition is the synthesis of *organic* materials with a high energy value from cell inputs or intakes of radiant or chemical *energy* and simple *inorganic* substances.

RADIANT SOLAR OR CHEMICAL ENERGY + SIMPLE INORGANIC SUBSTANCES: CO_2, H_2O AND NH_3 $\longrightarrow$ HIGH-ENERGY-VALUE ORGANIC COMPOUNDS: POTENTIAL ENERGY

(i) *Photosynthesis*

This is the formation of high-energy-value organic substances, using radiant or light energy (see Fig. 1.2), from inorganic carbon dioxide, water and ammonium compounds. It is the method of nutrition in all green plants and green bacteria in places where there is artificial or natural light. The important by-product is *oxygen*, which is essential for aerobic respiration.

(ii) *Chemosynthesis*

This is the formation of high-energy-value organic substances using energy from chemical processes and simple inorganic substances. Chemosynthesis is the method of nutrition of certain bacteria, in places where there is *no* light, as in caves, soil, and the depths of the ocean.

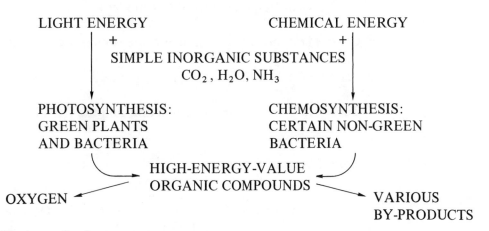

(b) Photosynthesis

The following chemical equation provides a summary of the reactants and products of photosynthesis. It does *not* show intermediate stages or metabolic pathways or the important intracellular enzymes controlling the process: these are summarised in 4.1(d).

$$6CO_2 \quad + \quad 12H_2O \quad + \quad LIGHT \longrightarrow C_6H_{12}O_6 \quad + 6O_2 + 6H_2O$$

CARBON   WATER   ENERGY   GLUCOSE   OXYGEN
DIOXIDE   (HIGH ENERGY
                                          VALUE)

Raw Materials for Photosynthesis

1. The *quality* of *light* (its wavelength, measured in nanometres, nm) plays a vital role in photosynthesis. Green plants can perform photosynthesis only when the available light is within the visible spectrum; they cannot utilise ultra-violet and infra-red. Red light (6000–7000 nm) and to a lesser extent blue light (4000–5000 nm) are best for green plants. The *energy content* of light is measured in joules, J; its brightness or intensity is measured in kilolux, klx. The *duration* of light received by a plant, i.e. the time of exposure and the day length, is also important in photosynthesis. Green plants show adaptations to variations in light quality, energy content and duration.
2. *Water* will vary in amount, and green plants show adaptations. *Hydrophytes* are water plants; *xerophytes* live in arid places, e.g. the desert; and *halophytes* grow in places with high salinity; *mesophytes* are the main type of green plants, and grow where the water supply is neither scanty nor excessive.
3. The *carbon dioxide* content of air is between 0.03% and 0.04%. It is the essential *inorganic* source of carbon.
4. *Chloroplasts* are the organelles and the place where photosynthesis occurs. They are energy-converting agents changing radiant energy into chemical energy stored in organic compounds.

$$\text{Radiant energy} \xrightarrow{\quad \text{chloroplasts} \quad} \text{Chemical energy stored in organic substances}$$

   Structure: Chloroplasts are mainly lens shaped, but in *Spirogyra* they have a spiral shape and in *Chlamydomonas* they are cup shaped. There are up to 100 in a single leaf mesophyll cell. There is a double *membrane*, and the pigment lies in *grana*; starch granules and lipid droplets are located in the *stroma*.

   Chlorophyll pigments: Chlorophyll *a* is the main functional pigment; others include chlorophyll *b*, carotenes, xanthophylls, and phycobilins. (See Section 15.4(a, 4).)

(c) Nutrient Requirements of Green Plants

Green plant nutrient requirements are summarised in Table 4.1.

✱ (d) Chemical Mechanism of Photosynthesis

Table 4.2 compares the Hill reaction and the Calvin reaction that occur in photosynthesis.

Total Chemical Process

$$6CO_2 + 12H_2O \xrightarrow[\substack{\text{CHLOROPHYLL} \\ \text{and INTRACELLULAR} \\ \text{ENZYMES}}]{\text{LIGHT}} C_6H_{12}O_6 + 6O_2 + 6H_2O$$

Also written:

$$6CO_2 + 6H_2O \longrightarrow C_6H_{12}O_6 + 6O_2$$

Table 4.1 A summary of the nutritional requirements of green plants

Chemical element and source	Function	Deficiency effect
Carbon from CO_2; hydrogen and oxygen from H_2O	Raw materials for photosynthesis	No formation of any organic material
Nitrogen from nitrates and ammonia	Protein and nucleic acid formation	Stunted growth; chlorosis (yellowing due to chlorophyll loss)
Phosphorus — as phosphates	ADP and ATP formation, proteins and cell membranes	Poor growth, dull green leaf colour
Potassium	Cell membrane formation	Leaf-edge yellowing and premature death of plant
Calcium from lime	Intercellular cell wall cement	Stunted root and stem growth
Iron	Chlorophyll formation	Chlorosis
Magnesium	Centre of chlorophyll molecule	Chlorosis

✱ **Table 4.2 Features of the light-dependent and light-independent stages of photosynthesis**

Light-dependent stage (Hill reaction) in in chloroplast grana	Light-independent stage (Calvin reaction) in chloroplast stroma
Light energy, absorbed by chlorophylls, is used as follows: (a) *ATP formation* (energy store) (b) *Photochemical* splitting of water into *hydrogen atoms* and *oxygen gas*, $2H_2O \longrightarrow 4H + O_2$	Carbon dioxide converted by hydrogen to the simple carbohydrate triose using ATP energy $3CO_2 + 12H \longrightarrow C_3H_6O_3 + 3H_2O$ TRIOSE Simple triose sugars are rapidly changed into glucose, $C_6H_{12}O_6$
Unaffected by temperature change; light *essential*; intracellular enzymes *not* involved	Affected by temperature change; light *not* essential; intracellular enzymes involved

(e) Leaf in Photosynthesis

1. There are four types of leaf in flowering plants:
    (a) *Cotyledons*, which are the seed leaves, are used mainly for food storage but can be photosynthetic in some plants.
    (b) *Scale leaves* are found on buds, where their function is protective; in an onion bulb they store food.
    (c) *Floral leaves* are known as *sepals*, and these are green and photosynthetic.
    (d) *Foliage leaves* are the main photosynthetic organs (Fig. 4.1).

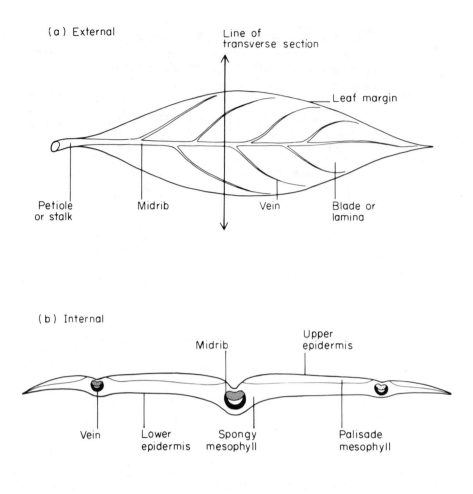

(a) External

Line of transverse section

Leaf margin

Petiole or stalk

Midrib

Vein

Blade or lamina

(b) Internal

Midrib

Upper epidermis

Vein

Lower epidermis

Spongy mesophyll

Palisade mesophyll

Figure 4.1   The structure of a dicotyledon leaf of privet, *Ligustrum* sp. In examinations you will probably be asked to 'label and explain these figures'.

Foliage Leaf Parts

  (i) Leaf base is the point of attachment to stem with *axillary bud*.
 (ii) Leaf stalk (*petiole*) is a continuation of stem structure.
(iii) Leaf blade (*lamina*) has *veins* as a continuation of vascular tissue.

Dicotyledon foliage leaves are *broad* with *net* veins; they have different upper and lower surfaces, and the upper surface is held at right angles to light.

Monocotyledon foliage leaves are *narrow* with *parallel* veins; upper and lower surfaces are similar, and the leaves grow upwards towards light.

2. The *internal structure of a dicotyledon foliage leaf* is illustrated in Fig. 4.2. A transverse section of a lamina shows the following:

  (a) *Epidermis*: upper and lower layers, with outer waxy *cuticle*.

  (b) *Guard cells* with chloroplasts, mainly in lower epidermis; two surround a pore for gas exchange (stoma). Stoma closed when guard cell *flaccid*, open when *turgid*. (See Fig. 4.3.)

  (c) *Mesophyll*: region between epidermal layers, consisting of loosely packed cells.

   (i) *Palisade* mesophyll, upper main photosynthesis region.

  (ii) *Spongy* mesophyll, lower main gas-exchange region via *air spaces*. In a *monocotyledon* leaf, mesophyll cells are all of one type.

66

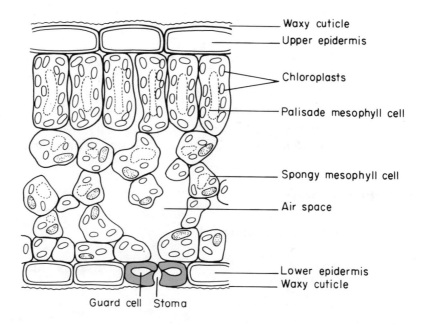

Figure 4.2   Internal structure of a dicotyledon leaf. In examinations you will probably be asked to 'label and explain this figure'.

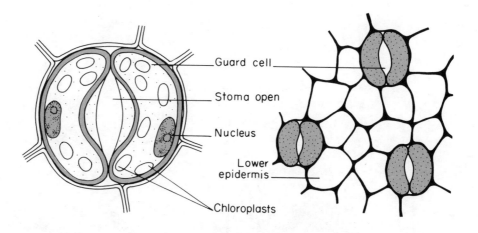

Figure 4.3   Surface view of air pores (stomata) of a dicotyledon leaf.

(d) *Veins* or *vascular bundles* for transport and support; a *bundle sheath* with fibres, chloroplasts and starch granules may surround the vein:
  (i) upper *xylem*
  (ii) lower *phloem*
3. The structure of the leaf enables it to perform the functions of solar radiant energy absorption and gas exchange.

Solar Radiant Energy Absorption

Abundant chloroplasts absorb radiant energy by means of chlorophyll *a* pigment, with enormous surface area:volume ratio.

Leaf lamina has large flat surface area, but is thin for ease of light penetration.

Leaf mosaic is the arrangement of leaves on the stem that allows the maximum exposure of the leaf surface to light in order to prevent *overshading*.

67

Gas Exchange

Abundant stomata allow rapid entry of CO_2 from air and exit of water vapour and oxygen gas.

Vein network consists of xylem tissue, an efficient transporter of *water* by means of the transpiration stream; and the phloem, which removes photosynthesis products and transports them around the plant.

Leaf air spaces allow rapid diffusion of gases and water vapour.

4.2 Practical Work

(a) Experimental work

Table 4.3 and Figs 4.4–4.7 summarise the methods used and the apparatus needed.

Figure 4.4 Apparatus to show that carbon dioxide is needed in photosynthesis.

Table 4.3 A summary of experiments on photosynthesis

Experiment	Method	Control
Starch test shows photosynthesis has occurred with formation of starch (Certain plants do not form starch.)*	1. Leaf immersed in boiling water 2. Warmed in ethanol for chlorophyll extraction 3. Softened in water 4. Treated with iodine solution 5. Blue-black colour in the parts with starch content	Destarching: keep potted plant (*Coleus*, geranium or nasturtium) in total darkness for 24 hours
Need for *carbon dioxide* (Fig. 4.4)	1. Destarched potted plant 2. Enclose one leaf in CO_2-free air over soda lime or potassium hydroxide 3. Expose to light for 6 hours and then perform starch test	1. Same plant 2. Enclose one leaf in air over water
Need for *light* (Fig. 4.5)	One leaf destarched; potted plant kept in light-proof flask for 6 hours and starch tested	Same plant; leaf in clear glass flask
Need for *chlorophyll* (see Section 15.4(a, 4) for pigment separation)	1. Trace pattern of *variegated* leaf of destarched potted plant 2. Expose to light 3. Starch test — positive where leaf areas were *green*	Leaf pattern tracing is control
Need for certain *mineral ions* (Fig. 4.6)	1. Prepare water culture solution defective in *one* chemical ion, either nitrate or iron 2. Observe growth effects 3. Chlorosis with defective photosynthesis	1. Grow seedling in *complete* culture solution 2. Healthy growth and normal green colour
Oxygen, a product of photosynthesis (Fig. 4.7)	1. Canadian pondweed kept under inverted glass filter funnel in 0.1% sodium hydrogen carbonate in well lighted position in a water bath maintained at 30°C for 3 days 2. Test gas collecting in inverted water-filled test tube for presence of oxygen with glowing splint	1. Maintain identical apparatus in *total darkness* for same period 2. Note the amount of gas that collects and test for oxygen
Rate of *photosynthesis* is also a measure of the *primary productivity*	1. Measured as the number of bubbles produced per minute from cut stem end of water weed 2. Measured as percentage increase in dry weight of leaf discs from same plant 3. Accurate measurement made in microlitres (μl) per hour per square centimetre (cm^2) of leaf	None Rate affected by: 1. *Temperature*: 10°C increase below 38°C doubles the rate of enzyme action (see Fig. 1.3(a)) (affects dark or Calvin reaction only 2. *Carbon dioxide* increases to an optimum concentration 3. *Light intensity* increases to an optimum value

*Certain plants, e.g. Iris, form only *glucose* in their leaves and no starch. In this case their juice is extracted and is tested with Benedict's or Clinistix reagent. No control is needed.

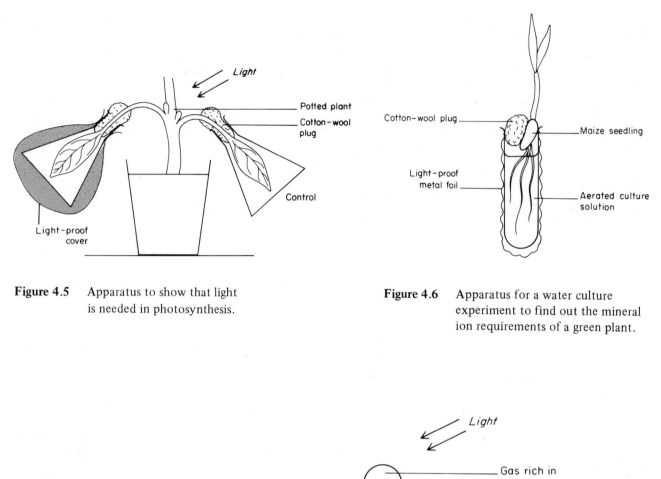

Figure 4.5 Apparatus to show that light is needed in photosynthesis.

Figure 4.6 Apparatus for a water culture experiment to find out the mineral ion requirements of a green plant.

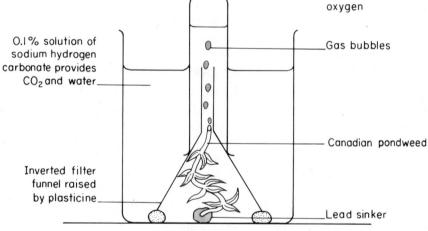

Figure 4.7 Apparatus to show gas production in photosynthesis using Canadian pondweed, *Elodea canadensis*.

(b) Limiting Factors

The rate of photosynthesis is governed by the factor in short supply (the limiting factor), if other factors are present in excess. The effects of changing temperature and light intensity on the rate of photosynthesis of an alga are shown in Fig. 4.8. Increasing temperature has no effect when light intensity is low, therefore light intensity is the limiting factor.

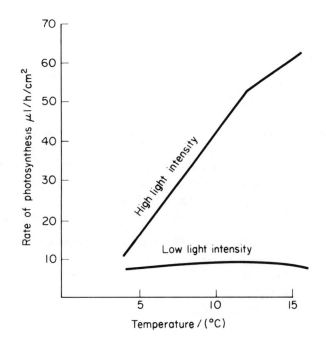

Figure 4.8 Graph to show the effect of temperature and light intensity on the rate of photosynthesis in an alga.

(c) Microscope Work

This involves the examination of prepared microscope slides showing transverse sections of a privet or similar leaf. *Electron micrographs* should also be viewed to recognise the chloroplast structure.

4.3 Examination Work

(a) Multiple-choice Objective Questions

1. The equation which best describes photosynthesis is:
   - (a) water + carbon dioxide + energy $\longrightarrow$ glucose + oxygen
   - (b) energy + carbon dioxide $\longrightarrow$ oxygen + glucose + water
   - (c) glucose + oxygen $\longrightarrow$ carbon dioxide + water + energy
   - (d) oxygen + water $\longrightarrow$ carbon dioxide + energy + glucose

2. In which one of the following is oxygen produced?
   - (a) a root hair cell
   - (b) a leaf epidermal cell
   - (c) a mesophyll cell
   - (d) a phloem cell

3. Which one of the following is *not* needed as a raw material by a green plant to make sugar by photosynthesis?
   - (a) water
   - (b) light
   - (c) oxygen from air
   - (d) chlorophyll

4. Which of the following releases oxygen during daylight?
   (a) a fungus
   (b) *Amoeba*
   (c) a germinating seed
   (d) *Spirogyra*

5. Why does grass turn yellow when covered by a wooden board?
   Because:
   (a) light is needed for photosynthesis
   (b) light is needed for chlorophyll formation
   (c) the wood does not contain iron
   (d) the board prevents oxygen reaching the grass

6. Magnesium is an important nutrient ion in green plants as it is essential as a component of:
   (a) cell sap
   (b) protein
   (c) chlorophyll
   (d) glucose

7. When a green plant performs photosynthesis at its maximum rate:
   (a) the rate of water loss is low
   (b) the water content of the plant will be low
   (c) the energy content of the plant will be high
   (d) the energy content will be unaffected

8. The yield of lettuce grown in a greenhouse can be increased by:
   (a) increasing the soil water content
   (b) increasing the air carbon dioxide content
   (c) increasing the air oxygen content
   (d) increasing the soil iron content

9. During the light-independent or dark reaction of photosynthesis the main process which occurs is:
   (a) release of oxygen
   (b) energy absorption by chlorophyll
   (c) adding of hydrogen to carbon dioxide
   (d) formation of ATP

10. Manufactured organic food formed in the leaves reaches the root by the:
    (a) leaf stomata
    (b) midrib phloem
    (c) midrib xylem
    (d) cambium

11. The purpose of warming a green leaf with denatured ethanol (methylated spirits) is to:
    (a) remove the leaf waxy cuticle
    (b) remove the green chlorophyll pigment
    (c) dissolve glucose and sucrose
    (d) expel oxygen gas from the leaf

12. The gas collected during photosynthesis of a submerged pondweed is a mixture in which the *major* component is:
    (a) oxygen
    (b) nitrogen
    (c) carbon dioxide
    (d) argon

13. Which of the following component cells of a leaf is *without* chloroplasts?
    (a) palisade mesophyll
    (b) spongy mesophyll
    (c) upper epidermis
    (d) guard cells

14. The method of nutrition or feeding of green algae is:
    (a) holozoic
    (b) heterotrophic
    (c) autotrophic
    (d) saprophytic

15. Green plants are unable to grow in depths of the ocean below 200 metres because of a lack of:
    (a) oxygen
    (b) carbon dioxide
    (c) light
    (d) mineral ions

16. The chemical reagent which will remove carbon dioxide and oxygen from air is:
    (a) iodine solution
    (b) potassium hydroxide
    (c) ethanol
    (d) alkaline pyrogallol

17. If a green plant was fed with water containing the heavy isotope tracer ^{18}O oxygen, the isotope would be finally located in:
    (a) the starch granules in a leaf cell
    (b) the oxygen formed by photosynthesis
    (c) the carbon dioxide formed in respiration
    (d) the cellulose of cell walls

18. The presence of ^{18}O oxygen heavy isotope is detected by means of:
    (a) a Geiger counter
    (b) an autoradiograph
    (c) a mass spectrograph
    (d) an electron micrograph

19. If a green plant is fed with the radioactive isotope carbon 14 as ^{14}C in carbon dioxide gas, it will be finally located in:
    (a) the water transpired from the leaf
    (b) the starch granules in the leaf mesophyll
    (c) the ash remaining on burning the plant
    (d) the nitrate formed on decay of the plant

20. Which of the following ranges of temperature increase would double the rate of photosynthesis in a green plant with an ample supply of water, light and carbon dioxide?
    (a) 45–55°C
    (b) 25–35°C
    (c) 30–35°C
    (d) 20–25°C

(b) Structured Questions

Question 4.1

(a) Name two environmental factors, other than light, which may affect photosynthesis. For each factor you name, explain why its absence may prevent photosynthesis occurring.

(b) In what form is the carbohydrate produced during photosynthesis
   (i) transported through a potato plant?
   (ii) stored underground in a potato tuber?

(c) The accompanying drawing shows a potato tuber. Potatoes are often called root crops. Label two structures on the drawing which show it is in fact a stem. **(6 marks)**
(AEB, 1982)

∗ Question 4.2

A large glass-stoppered flask containing *Nasturtium* leaves and air enriched with carbon dioxide was exposed to bright sunlight for 3 hours. The diagram represents the apparatus used. Two 100 cm³ samples of gas were withdrawn from the apparatus. The first (sample A)

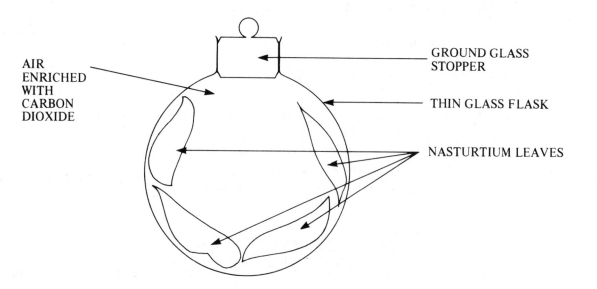

was withdrawn at the beginning of the experiment and the second (sample B) was withdrawn at the end of the 3 hours. The volume of each sample was measured after exposure, firstly to potassium hydroxide and secondly to alkaline pyrogallol solutions. The results are tabulated below.

	First sample, A	Second sample, B
Original volume	100 cm^3	100 cm^3
Volume after exposure to potassium hydroxide solution	94 cm^3	98 cm^3
Volume after exposure to alkaline pyrogallol solution	73 cm^3	73 cm^3

(i) Name the gas absorbed by the potassium hydroxide solution and calculate the percentage of the gas present in each of the samples A and B.

(ii) Name the gas absorbed by the pyrogallol solution and calculate the percentage of this gas present in each of the samples A and B.

(iii) Name the biological process which brought about the changes in the composition of the gases.

(iv) Why was the flask exposed to bright sunlight?

(v) Why was air enriched with carbon dioxide used in the experiment?

(vi) Describe briefly the other process in the leaves which affects these results.　**(14 marks)**

(NISEC)

(c) Free-response-type Questions

(i) *Short-answer Questions*

1. Draw a labelled diagram of a *stoma* and a few of the surrounding cells as seen in a *vertical section*. Show the distribution of chloroplasts in the diagram.
(6 marks)

＊ 2. Describe an experiment to show that chlorophyll is necessary for the production of starch in a leaf. Include details of a chemical test for starch. **(11 marks)**

3. State the percentage volume of carbon dioxide in the air. How is this maintained?　**(5 marks)**

4. Enumerate the main stages in the experimental testing of a green leaf for the presence of starch.　**(5 marks)**

(ii) *Long-answer Questions*

＊ **Question 4.3**

Give an illustrated account of the *shape* and *internal structure* of the leaf of a flowering plant and explain how the leaf is *suited* to photosynthesis.　**(25 marks)**

(L)

Answer to Q4.3

This is a straightforward or *direct* question based entirely on the topic of photosynthesis. The answer requires *three* labelled diagrams:

1. Leaf shape or form as for Fig. 4.1(a).
2. Internal structure of leaf shown in a *general* diagram, Fig. 4.1(b).
3. Detailed structural diagram as in Fig. 4.2.

The leaf's suitability to photosynthesis is listed as follows:

1. Thin and large surface area directed to receive maximum light.
2. Porous epidermis for entry of carbon dioxide and exit of oxygen and water.
3. Leaf mosaic to prevent overshading, and leaf arrangement on stem – called phyllotaxy.
4. Enormous number of chloroplasts in mesophyll: site for photosynthesis with large surface area : volume ratio.

* **Question 4.4**

    (a) List *four* features of a leaf which show that it is able to carry out photosynthesis efficiently.

    (b) How would you show experimentally that light is necessary for photosynthesis?

    (c) State two other factors, apart from light, which are necessary for photosynthesis to occur in a leaf.

(20 marks)
(SUJB)

Answer to Q4.4

A *direct* question.

(a) Leaf features for efficient photosynthesis:

    1. Thin, porous structure for entry of carbon dioxide and exit of oxygen and water vapour.
    2. Leaf arrangement at right angles to light in dicotyledons and upwards towards the light in monocotyledons.
    3. Enormous surface area : volume ratio of chloroplasts together with their great number in mesophyll as site for photosynthesis.
    4. Leaf surface area forms the greatest part of the plant body.

(b) To show that light is necessary for photosynthesis is a question on practical work, and the apparatus is described in Fig. 4.5.

    Important points: to use destarched potted plant, and also to describe 'control' experiment kept in total darkness. It is also necessary to outline the iodine solution test for formation of starch as *evidence* that photosynthesis has occurred.

(c) Since the word *state* is used, you are expected simply to name *two* other factors; carbon dioxide supply, water supply, suitable temperature (essential for 'dark reaction') or supply of essential mineral ion, e.g. magnesium to form chlorophyll, and the presence of chlorophyll itself.

* **Question 4.5**

    (a) What is photosynthesis?   **(4 marks)**

    (b) Outline the Hill (or light) reaction and the Calvin (or dark) reaction.   **(6 marks)**

    (c) How would you show experimentally:
        (i) that a plant needs light to carry out photosynthesis?   **(8 marks)**
        (ii) that the rate of photosynthesis varies with light intensity?   **(5 marks)**
        (iii) What factors eventually limit the rate of photosynthesis?   **(2 marks)**

(OLE)

Answer to Q4.5

3. A *direct* question.
   (a) *Definition* of photosynthesis: process in green plants of producing organic material with high energy content from simple inorganic materials, water and CO_2, using radiant sunlight energy absorbed by chloroplast pigments; the important by-product is oxygen.

   This definition could be summarised using the equation:

$$6CO_2 \ + \ 6H_2O \ + \ \overset{\text{Radiant}}{\underset{\text{energy}}{}} \xrightarrow{\text{CHLOROPLASTS}} C_6H_{12}O_6 \ + \ 6O_2$$

   Carbon   water   glucose   oxygen
   dioxide

   (b) *Hill or light reaction outline.* Light essential.
      (i) radiant energy absorption by chloroplasts
     (ii) formation of ATP energy store
    (iii) *photochemical* reaction unaffected by temperature to produce hydrogen atoms and oxygen gas:

$$4H_2O \longrightarrow 8H + 2O_2$$

   Calvin or dark reaction outline. Light not needed.
      (i) *Thermochemical* reaction called *reduction* of carbon dioxide by hydrogen atoms to form triose sugars.
     (ii) *Enzyme* controlled and therefore affected by temperature.
    (iii) Glucose rapidly formed from triose.
    (iv) Starch rapidly formed from glucose by condensation.

   (c) This requires an experiment which can also demonstrate effect of light intensity on the rate of photosynthesis.
      (i) The Canadian pondweed apparatus shown in Fig. 4.7 demonstrates how oxygen is a product of photosynthesis and therefore is evidence that photosynthesis has occurred. The plant and apparatus are maintained for up to 48 hours in total darkness. Little if any gas collects and is insufficient to relight a glowing wood splint, which would have indicated presence of some oxygen.
     (ii) Light intensity from the same source, say an 80-watt table lamp, will decrease with distance of the lamp from pondweed apparatus. Since lamp also radiates *heat*, it is necessary to place a glass tank of water between lamp and the apparatus to act as a heat filter.

   Erect *two* sets of pondweed apparatus, one set at a distance which is *half* that between the other set and lamp. Comparison of rates can be determined by a bubble count from cut weed stem surface per hour. Weed apparatus nearer the lamp shows the most rapid rate, compared with the apparatus furthest away from the lamp.
    (iii) The factor which eventually limits the rate is the *limiting factor* or the one present in smallest amount.

4.4 Self-test Answers to Objective and Structured Questions

Answers to Multiple-choice Objective Questions

1. a   2. c   3. c   4. d   5. b   6. c   7. c   8. b   9. c   10. b   11. b   12. b
13. c   14. c   15. c   16. b   17. b   18. c   19. b   20. b

Answer to Structured Question 4.1

(a) Factor 1: water
    Reason: Essential raw material for photosynthesis
    Factor 2: carbon dioxide
    Reason: essential raw material for photosynthesis

(b) This section involves a knowledge of transport (Chapter 6).
    (i) Form of carbohydrate transported through a potato plant is sucrose (disaccharide).
    (ii) Form of carbohydrate stored underground in a potato tuber is starch (polysaccharide).

(c) Stems are recognised by the presence of a leaf, reduced in the potato tuber to a leaf scar with a bud in the leaf scar axil. These together form the potato 'eyes'. Additional features indicating it is a modified stem include the adventitious roots arising from the base of the developing shoot.

Answer to Structured Question 4.2 (see also Section 7.2b)

(i) Gas absorbed by potassium hydroxide is *carbon dioxide*.

$$\% \text{ in sample A} = \frac{100 - 94}{100} = \frac{6}{100} = 6\%$$

$$\% \text{ in sample B} = \frac{100 - 98}{100} = \frac{2}{100} = 2\%$$

(ii) Gas absorbed by alkaline pyrogallol solution is *oxygen*.

$$\% \text{ oxygen in sample A} = \frac{100 - 73}{100} = \frac{27}{100} = 27\% - 6\% \, CO_2 = 21\% \, O_2$$

$$\% \text{ oxygen in sample B} = \frac{100 - 73}{100} = \frac{27}{100} = 27\% - 2\% \, CO_2 = 25\% \, O_2$$

(iii) Biological process – photosynthesis.

(iv) Flask exposed to bright sunlight to obtain the *radiant energy* input essential for photosynthesis.

(v) Carbon dioxide enriched air provides a source of one of the raw materials needed for photosynthesis, namely carbon dioxide.

(vi) The other process which affects these results is *aerobic respiration*, which involves an intake of oxygen (see Chapter 7).

5 Animal Nutrition

5.1 Theoretical Work Summary

(a) Heterotrophic Nutrition

Heterotrophic nutrition is the input or intake of high-energy-value organic substances (food) by the cells of animals, fungi and certain bacteria, which are described as heterotrophic organisms (Table 5.1).

Table 5.1   Heterotrophic organisms and their food sources

Organism	Organic food source
A. Consumers:	
(i) *Herbivores*: zooplankton, caterpillars, locust, rabbit	Autotrophic plants
(ii) *Carnivores*: beetles, sharks, cats and seals	Other heterotrophic animals
(iii) *Omnivores*: human beings	Autotrophic plants and heterotrophic animals
(iv) *Detritivores*: earthworms, tubeworms, woodlice, millipedes	Dead plant material
B. Decomposers: bacteria and fungi	Break down organic materials made by plants and animals — decay organisms

(b) Food Composition

1. Foods are mixtures of different chemical compounds, mainly organic, and some inorganic, collectively called *food nutrients*, which have certain functions in the animal body. *Food energy value* is measured in kilojoules (kJ) by means of *calorimeters*. (See Table 5.2.)
2. Food functions to provide animal body cells with:
   (i) *energy* for cell metabolism, work, movement, warmth and glandular activity;
   (ii) *structural materials* for cell growth and repair;
   (iii) *functional materials* for cell metabolism and cell protection.
3. *Plant*-origin foods are rich in dietary fibre, vitamins C and K and folic acid. *Animal*-origin foods are rich in vitamins A, D and B_{12}, and essential amino acids.

(c) Dietary Intake in Human Beings

1. *Omnivores*, such as human beings, feed on a *mixed diet* of food from plant and animal sources.
   (a) *Strict vegan or herbivore diet* for humans is *deficient* in certain essential amino acids, vitamin B_{12} and calcium.

79

Table 5.2 Role and energy values of food nutrients

Food nutrient	Energy provision	Structural purpose	Functional purpose
CARBOHYDRATES	Provide 17 kJ/g + heat	None (see * below)	Glucose in all living cells. Glycogen store in muscle and liver
LIPIDS	Provide 38 kJ/g + heat	Adipose tissue. Skin and hair formation — essential fatty acids	Heat insulation
PROTEINS	*Can* provide 17 kJ/g; is a *wasteful* use of protein	Muscle and connective tissue: collagen, myosin, elastin and keratin. (*Mucopolysac-charides formed of protein + carbohydrates chitin, chondroitin of exoskeleton and connective tissues.)	Enzymes, hormones, anti-bodies
ETHANOL (not essential nutrient)	*Can* provide 29 kJ/g	None	None. Can cause alcoholic *addiction*
MINERAL IONS	(P) ~ ATP formation ~ high-energy phosphate bond	Ca: teeth and bones P: teeth and bones F: teeth and bones Fe: blood and muscle	Ca: muscle contraction, blood clotting Fe: cell respiration I: thyroxine hormone NaCl: nerve activity, gastric juice
VITAMINS	B_1, B_2 and nicotinic acid: energy release	A: night vision D: teeth and bones B_{12} and folic acid: red blood cells C: connective tissues	A: disease resistance K: blood clotting
WATER	None	Cell turgidity — support	Temperature control. Solvent for metabolic changes
DIETARY FIBRE	None (provides 17 kJ/g in *herbivores*.)	None	Healthy bowel activity in peristalsis

(b) *Strict carnivore diet* for humans is deficient in vitamin K and dietary fibre.

2. *Daily dietary intake* has the following general composition based on percentage of energy needs.

	Protein	Carbohydrate	Lipids
High dietary fibre — poor developing countries	10%	80%	10%
Low dietary fibre — wealthy developed countries	10%	55%	35%

3. *Daily energy needs* measured in *megajoules* (MJ) vary according to a person's *age, work* or *pastime* activity, and whether the female is *pregnant* or *lactating*.

Age in years	<1	Up to 3	Up to 6	Up to 9	Boys 10–19	Girls 10–19	Men	Women	Pregnant	Breast-feeding
Megajoules (MJ)	4	6	8	10	11–12	9–10	12	9	10.5	11.5

4. Daily needs of *minerals and vitamins*: the *small* amounts, *micrograms* (μg), of vitamins A, D, B_{12} and folic acid, and *milligrams* (mg), of *all* other vitamins and minerals will naturally be available provided a *mixed diet* of the correct *energy composition* (section 5.1(c)) is taken.
5. Food additives are chemical substances added to *processed* foods to restore flavour, colour or texture. Tartrazine dye, a yellow food colour, is such an additive, and causes allergy in certain people.

(d) Dietary Balance

1. *Balanced diets* are composed of *daily* recommended intakes or *optimum* amounts of energy from carbohydrate and lipids, together with protein, vitamins, mineral ions and dietary fibre (at least 10 g daily) and sufficient water, to maintain good health.
2. *Malnutrition* or dietary *imbalance* arises from either a deficiency or an excess intake of nutrients, or undernutrition or overnutrition. (See Table 5.3.)

Table 5.3 The importance of a balanced diet: one without dietary excess or deficiency

Nutrient	Undernutrition: dietary deficiency	Overnutrition: dietary excess
ENERGY AND PROTEIN	*Starvation* – all food and nutrients deficient; *marasmus* – child starvation; *kwashiorkor* – protein deficiency, but adequate carbohydrate intake	*Obesity* and overweight: energy intake *exceeds* energy output; overeating; lack of exercise. Factor in heart disease, bowel disorders, sugar diabetes, gout, joint disorders
VITAMINS – LIPID SOLUBLE		
A	*Night blindness* and *keratomalacia* (eye disorder)	*Hypervitaminosis*: excess intake causes bone and muscle pains
D	*Rickets* in children, *osteomalacia* in adults, bone softening and defective teeth	As above
VITAMINS – WATER SOLUBLE		
B_1	*Beri-beri* affecting heart and nerves	None – appetite stimulant
B_2	*Pellagra*, skin, nervous, and digestive disorders	As above
B_{12}	*Pernicious anaemia* affecting certain red blood cells	None – water soluble therefore excess excreted in urine
Folic acid	Type of *anaemia*	None – see above
C	*Scurvy* – bleeding, swollen gums and internal bleeding	None – see above
MINERALS		
Calcium	*Osteoporosis* (bone softening)	Not harmful
Iron	*Nutritional anaemia* affecting red blood cells	*Siderosis*, iron collecting in liver and bone
Sodium	*Dehydration* and muscular cramps, mental apathy	*Harmful* in babies. Cow's milk has more than human milk. Factor in *heart disease*
Iodine	*Goitre*, thyroid gland enlargement	Harmful and poisonous
Fluoride	Tooth *decay*	*Fluorosis* or enamel discoloration
DIETARY FIBRE	*Constipation*; diseases linked include piles, bowel cancer, appendicitis, gallstones, heart disease	*Laxative*, bowel toner. *Phytic acid* content affects absorption of calcium, zinc and iron
WATER	Dehydration	Water intoxication (rare condition)

(e) Heterotroph Feeding Methods

Table 5.4 summarises the methods used by heterotrophs for ingestion or to capture food. These methods are illustrated in Figs 5.1–6.

Table 5.4 Ingestion: food capture and feeding methods in heterotrophs

Food's physical form	*Method, structure and organism*
A. SMALL PARTICLES in suspension in water; feeding usually *continuous*	1. *Filtration*: *gill rakers* (strain plankton), e.g. herring; *baleen* is an upper jaw comb which strains krill (in baleen whales) 2. *Cilia*: cilia on body e.g. in *Paramoecium* on *gill filaments*; bivalve molluscs filter out food particles 3. *Pseudopodia*: engulf food particle in *food vacuole* as in *Amoeba*

B. LARGE PARTICLES of plant or animal material (Fig. 5.2)	1. *Not broken up (without teeth)* and *swallowed whole*, e.g. coelenterates, earthworms, dogfish, snakes, birds, anteaters
	 2. *Broken up with teeth*: most mammals (Figs 5.5 and 5.6)
	 3. *Broken up*:
	   (a) *with mandible* – crustaceans, insects (Fig. 5.2), centipedes, millipedes
	   (b) *with radula* – snails and many other molluscs

C. FLUIDS, i.e. blood, nectar, phloem, sap, gut fluid (Figs 5.3, 5.4)	1. *Surface absorption*: endoparasites in gut, e.g. tapeworm; in blood, e.g. malarial parasite 2. *Sucking*: butterflies (nectar) (Fig. 5.4); *male* mosquito; after external digestion, e.g. housefly (Fig. 5.3) 3. *Piercing and sucking*: blood collected by *female* mosquito (Fig. 5.1) and by fleas; greenfly (*Aphis*) collects plant juices from vascular tissue

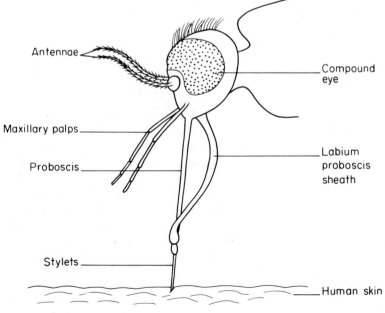

Figure 5.1 Head of a female mosquito, *Anopheles* sp.

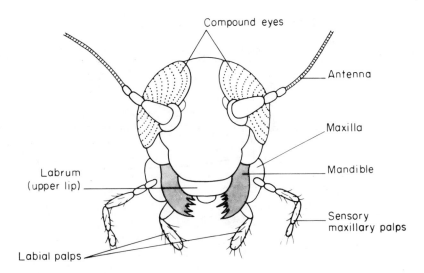
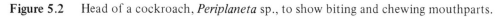

Figure 5.2   Head of a cockroach, *Periplaneta* sp., to show biting and chewing mouthparts.

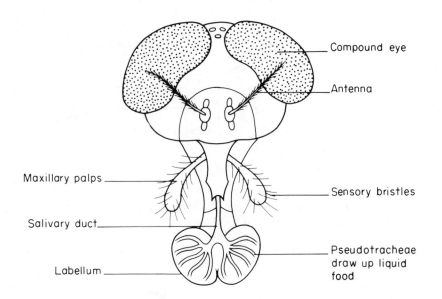

Figure 5.3   Head of a housefly, *Musca domestica*, to show sucking mouthparts.

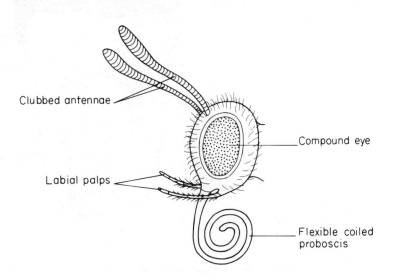

Figure 5.4   Head of a butterfly, *Pieris* sp., to show sucking mouthparts.

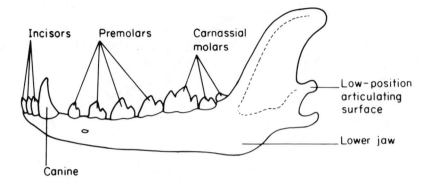

(a) Dog

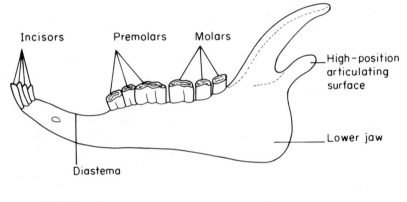

(b) Sheep

Figure 5.5   Lower jaw and teeth of (a) a typical carnivore, a dog, *Canis familiaris*; and (b) a typical herbivore, a sheep, *Ovis* sp.

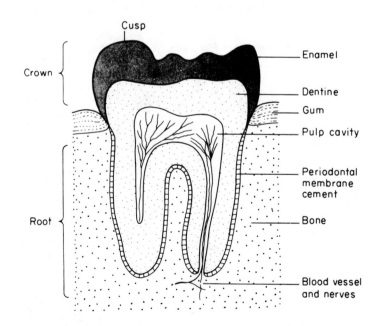

Figure 5.6   Internal structure of a human molar tooth seen in vertical section. In examinations you will probably be asked to 'label and explain this figure'.

(f) Digestion

Digestion is the conversion of *large*, mainly insoluble, complex, organic molecules of food into simpler, *smaller*, soluble molecules of food nutrients by the combined processes of:

1. *physical digestion*: to increase food surface area, i.e. blending, melting, dissolving, emulsification and lubrication;
2. *chemical digestion*: mainly *hydrolysis* (reaction with water), in the presence of *extracellular enzymes*, controlled by *hormones* and *pH*. (*Cooking* alters food physically and chemically.)

    In *extracellular digestion*, the digestive enzymes are released *on to* food:
    (a) *in* gut cavity — mammals;
    (b) *outside* body — housefly saliva.

These extracellular enzymes are called *carbohydrases* for carbohydrates, *lipases* for lipids, *proteinases* for proteins, and *nucleases* for nucleic acids.

(g) Summary of Chemical Digestion in Human Beings

Food nutrients not digested include: water, mineral ions, vitamins, ethanol (alcohol), propanetriol (glycerol), simple monosaccharides, glucose and fructose, and amino acids (often components of soups, gravies and stews). *Dietary fibre* (cellulose, lignin, pectins and gums) is not, or only partly, digested — previously called roughage or bulk.

    Foods digested include: carbohydrates (except monosaccharides), lipids, proteins and nucleic acids. (See Table 5.5.)

    In *intracellular digestion* cells *engulf* food material by:
    (a) *phagocytosis* of mainly solids — as in *Amoeba*, white blood cells, and intestinal epithelia (*microvilli* brush border).
    (b) *pinocytosis* of mainly fluids — the ingested food is then digested within the cell by intracellular enzymes secreted by the organelles known as *lysosomes*.

Mammal Gut

The structure of the alimentary canals of an *omnivore* (human being) and a *herbivore* (rabbit) are shown in Figs 5.7 and 5.8. Many herbivores digest *cellulose* by means of the enzyme *cellulase* produced by gut micro-organisms present in the caecum.

    Peristalsis (see also page 191) is the method by which food moves through the gut aided by its dietary fibre content.

(h) Absorption of Nutrients

1. *Epithelial lining* of the gut from mouth to anus is absorptive and thin — one cell layer — and continuously moistened by mucus.
2. *Absorption* into thin, moist, epithelial cells is by diffusion, osmosis, active transport, pinocytosis or phagocytosis.
3. *Large surface area* of gut for absorption increased by stomach infoldings; or small intestine finger-like projections, *villi*, and *epithelial microvilli* (Fig. 5.9). The fate of the absorbed nutrients is summarised in Table 5.6.

Table 5.5 Summary of chemical digestion in the human gut

Secretion	Nutrient, enzyme, pH and products			
	Carbohydrate	Proteins	Lipids	Nucleic acids
SALIVA: salivary glands in	Starch (cooked) —amylase→ maltose pH 7.3 (+ trace glucose) (slightly alkaline)	Nil	Nil	Nil
GASTRIC JUICE: stomach wall gastric glands, 2–3 litres daily + mucus	Nil	Protein —pepsins→ peptone + peptides pH 1 HCl (strong acid) Milk: soluble caseinogen —rennin→ Insoluble curds (casein) (children only)	Lipids —lipase→ Alkanoic (fatty) acids + propanetriol (glycerol) (children only)	Nil
BILE JUICE: liver, gall bladder into duodenum, 500 ml daily	Nil	Nil	Lipids —bile cholic acid salts→ Emulsion of droplets pH 7–8 (alkaline)	Nil
PANCREATIC JUICE: pancreas into duodenum, 1500 ml daily	Starch —amylase→ maltose pH 7–8 (alkaline)	Protein peptone —chymotrypsin and trypsin→ Peptides + amino acids pH 7–8 (slightly alkaline)	Lipids —Lipase→ Alkanoic (fatty) acids + propanetriol (glycerol) pH 7–8 (alkaline)	Nucleic acids —nuclease→ nucleotides pH 7–8 (alkaline)
INTESTINAL JUICE: glands in small intestine wall + mucus	Maltase, sucrase and lactase Maltose Sucrose Lactose —disaccharidases→ monosaccharides pH 7–8 (alkaline)	Peptides —peptidases (erepsin)→ amino acids pH 7–8 (slightly alkaline)	Nil to *small* amount lipase	Nucleotides —nucleotidases→ nitrogen bases + pentoses + phosphate
LARGE INTESTINE: mucus from mucous gland	Nil	Nil	Nil	Nil

*

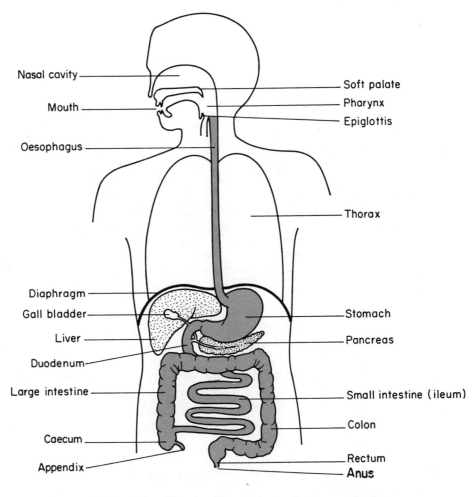

Figure 5.7 Human alimentary canal. In examinations you will probably be asked to 'label and explain this figure'.

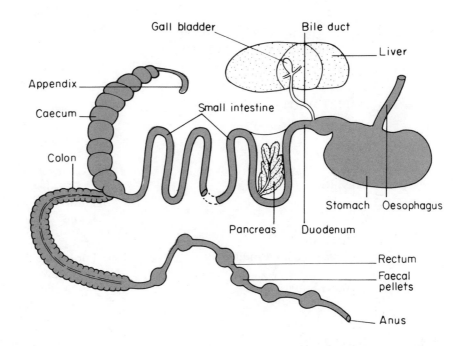

Figure 5.8 The main regions of a mammalian herbivore gut (rabbit, *Oryctolagus* sp.).

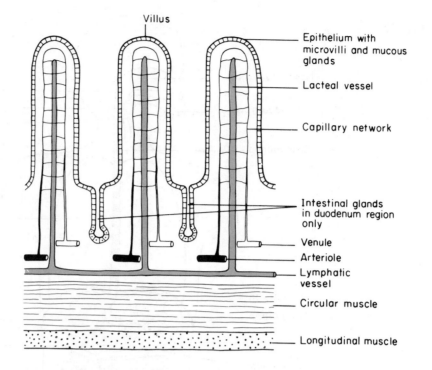

Villus

Epithelium with microvilli and mucous glands

Lacteal vessel

Capillary network

Intestinal glands in duodenum region only

Venule

Arteriole

Lymphatic vessel

Circular muscle

Longitudinal muscle

Figure 5.9 Diagram to show the structure of the wall of the small intestine.

Table 5.6 Fate of absorbed major nutrients — assimilation

Nutrient	Fate
GLUCOSE	To *tissue fluid* bathing all cells (see Fig. 6.4), enters cells aided by *insulin* hormone. Surplus glucose circulates in blood plasma and is converted to *glycogen*, which is stored in liver and skeletal muscles. *Obesity* occurs when glucose intake and lipid formation are excessive
ALKANOIC (FATTY) ACIDS AND PROPANETRIOL (GLYCEROL)	To *tissue fluid* for cell use. Surplus recombines and forms lipids in fat beneath skin and around kidneys. *Obesity* occurs when lipid intake is high
AMINO ACIDS	To *tissue fluid* to make protein for cell use. Surplus is *deaminated* in liver to form urea (carbamide) and glucose, which in turn can be changed into lipids (leading to obesity) and glycogen
MINERALS	
Calcium	Forms teeth and bones
Iron	Forms blood. Surplus stored in liver
VITAMINS	
Lipid soluble (A and D)	For cell use. Surplus stored in adipose tissue and liver
Water soluble	B_{12} and folic acid stored in liver. Surplus vitamins of B group and vitamin C excreted in urine
WATER	Is the important *solvent* for all metabolic activities and component of blood plasma.

5.2 Practical Work

(a) Food Tests

Samples of food can be tested for carbohydrates, lipids and proteins by the chemical tests given in Table 1.2.

(b) Extracellular Enzymes in Digestion

Experiments to demonstrate the roles of extracellular enzymes in digestion are summarised in Table 5.7. The enzyme and substrate solutions are placed in small test tubes and maintained in a water bath.

Table 5.7   Experiments to demonstrate the role of extracellular enzymes in digestion

	Amylase	*Pepsin*	*Trypsin*
Enzyme	Collect saliva from previously rinsed mouth. Test for absence of reducing sugars	Use pepsin powder	Use trypsin powder from beef pancreas
Substrate	Non-sweetened oatcake, cream cracker. Check for presence of reducing sugars	Thin strips of boiled egg white. Coagulated egg white drawn into glass capillary tubes	Strips of photograph negative (*gelatine* on plastic film)
Normal pH	Neutral (pH 7)	Make acid by adding 3 drops HCl	Neutral pH 7
Vary pH	By making acid with HCl, or alkaline with sodium hydrogen carbonate	Same	Same
Normal temperature	Maintain water bath at 37°C	Same	Same
Vary temperature	0°C in ice water; 100°C in boiling water	Same	Same. (Note gelatine dissolves in boiling water.)
Control	No enzyme	No enzyme	No enzyme
Evidence of digestion	Presence of reducing sugar shown by Benedict's or 'Clinistix' test reagent strip (specific for glucose)	Disappearance of solid white egg albumen	Dissolving of gelatine

CARE! All apparatus should be thoroughly cleansed and sterilised if it has been used to contain saliva.

(c) Passage of Digested Food through a Membrane

1. Fill a dialysis tube (visking membrane) with 5% starch and glucose mixture solution. Rinse off exterior with fresh tap water.
2. Immerse dialysis tube in distilled water.
3. Test distilled water for glucose and starch after 15 minutes.
4. Glucose only passes by *diffusion* into distilled water from dialysis tube.

5.3 Examination Work

(a) Multiple-choice Objective Questions

1. Rickets is a deficiency disorder due to a shortage of the vitamin:
   - (a) A
   - (b) B_1
   - (c) D
   - (d) C

2. The food substance which is digested in a rabbit but *not* in human beings is:
   - (a) glucose
   - (b) starch
   - (c) cellulose
   - (d) sucrose

3. The colon is a component part of a mammal's:
   - (a) stomach
   - (b) small intestine
   - (c) oesophagus
   - (d) large intestine

4. Gastric juice is a substance produced by or is a secretion of the:
   - (a) salivary glands
   - (b) liver
   - (c) stomach wall
   - (d) intestinal wall

5. Which of the following food substances, when surplus to the body needs, undergoes conversion by deamination?
   - (a) glucose
   - (b) propanetriol (glycerol)
   - (c) amino acids
   - (d) fatty (alkanoic) acids

6. The process of deamination occurs in the:
   - (a) kidney
   - (b) liver
   - (c) large intestine
   - (d) stomach

7. Poor vision in dim light or at night-time is due to a deficiency of vitamin:
   - (a) A
   - (b) B_1
   - (c) C
   - (d) D

8. The *main* enzyme present in the secretion from the salivary glands of human beings is:
   - (a) amylase
   - (b) lipase
   - (c) protease
   - (d) sucrase

9. Which of the following components of bile juice has a function in digestion?
   - (a) bile pigments
   - (b) cholesterol

(c) bile salts and acids

(d) urea (carbamide)

10. Which one of the following food substances is essential for the growth and repair of worn-out body tissues?
    (a) proteins
    (b) carbohydrates
    (c) lipids
    (d) dietary fibre

11. Which one of the following substances is not a waste product of cell metabolism?
    (a) urea (carbamide)
    (b) faeces
    (c) carbon dioxide
    (d) uric acid

12. Which of the following nutrients *cannot* be stored in large amounts in the human body?
    (a) lipids
    (b) iron
    (c) vitamin C
    (d) vitamin A

13. The dentition of a herbivorous animal differs from that of a carnivore in having *no*:
    (a) incisors
    (b) canines
    (c) premolars
    (d) molars

14. A component of bile juice which alters pH functions in the:
    (a) large intestine
    (b) stomach
    (c) oesophagus
    (d) small intestine

15. The secretions of the pancreas and liver enter the mammal alimentary canal at one of the following parts:
    (a) duodenum
    (b) rectum
    (c) stomach
    (d) colon

16. The percentage by weight composition of a food is as follows: carbohydrate 4.7%, protein 3.3%, lipids 3.8%, dietary fibre 0%, and water 88.2%. This food is most likely to represent the composition of one of the following foods:
    (a) roast beef
    (b) milk
    (c) potatoes
    (d) wholewheat bread

17. An enzyme component of pancreatic juice changes starch into maltose; this enzyme is called:
    (a) nuclease
    (b) trypsinogen
    (c) lipase
    (d) amylase

18. The small intestine is an efficient organ for absorption of digested food because it:
    (a) has a good nerve supply
    (b) has muscular walls for peristalsis
    (c) has intestinal folds and villi
    (d) is of short length

19. Filtered fruit juices are a rich source of which one of the following carbohydrates?
    (a) starch
    (b) cellulose
    (c) glycogen
    (d) glucose

20. Which of the following is an important form of storage of carbohydrate in living mammal muscle and liver?
    (a) cellulose
    (b) starch
    (c) glycogen
    (d) sucrose

(b) Structured Questions

Question 5.1

The table below shows the composition of five foods.

Food	Kilojoules per 100 g	Protein	Fat	Carbohydrate	Vitamin C	Vitamin D	Iron
				Composition per 100 g			
A	3800	0.4 g	86 g	0	0	40 mg	0
B	130	1.2 g	0	8 g	220 mg	0	0
C	1150	8.8 g	1.5 g	60 g	0	0	0
D	400	2.0 g	0.1 g	25 g	10 mg	0	6 mg
E	1650	0	0	100 g	0	0	0

From the table above select the food which would
(a) help to prevent anaemia
(b) provide most energy
(c) be best for growth
(d) consist only of sugar (sucrose)
(e) be best to prevent rickets
(f) be best for someone suffering from scurvy

(6 marks)
(NEA)

Question 5.2

(a) The diagrams show parts of the digestive systems of a herbivore and of a human being (not drawn to the same scale).

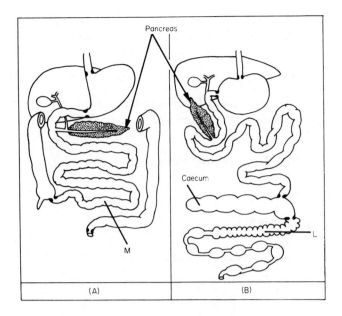

(i) Which diagram, A or B, shows the human digestive system? **(1 mark)**

(ii) Give the name, and state one function, of parts L and M. **(2 marks)**

(iii) Explain how the function of the caecum is related to the diet of B. **(2 marks)**

(iv) What function, apart from the production of insulin, is carried out by the pancreas? **(1 mark)**

(b) The diagram below shows the skull of a carnivore, together with its dental formula.

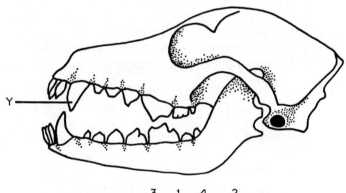

Dental formula: $i\frac{3}{3}$ $c\frac{1}{1}$ $p\frac{4}{4}$ $m\frac{2}{3}$

(i) Name the type of tooth labelled Y. **(1 mark)**

(ii) What is the total number of teeth possessed by this animal? **(1 marks)**

(SEB)

(c) Free-response-type Questions

(i) *Short-answer Questions*

1. (i) Name two digestive enzymes found in mammals, and name the glands which secrete them. **(2 marks)**

(ii) State the chemical reactions they help to bring about. **(2 marks)**

(SUJB)

2. Describe the main steps in the digestion and absorption of fats in a named mammal. What happens to the absorbed fats? **(8 marks)**

(NISEC)

3. What are the four kinds of teeth found in most mammals called?

4. Give one good food source, and an example of one symptom of the deficiency of vitamin A, vitamin D and vitamin C.

(ii) *Long-answer Questions*

＊ **Question 5.3**

Give an account of the digestion of *carbohydrates* in a named *herbivorous* mammal.

(25 marks)

(L)

Answer to Q5.3

A *direct* question seeking a description of chemical digestion of carbohydrates with names of enzymes, products and pH conditions. See Table 5.5.

Layout

1. Mouth – amylase digestion
2. Stomach – nil digestion
3. Bile juice action – nil digestion
4. Pancreatic juice digestive action in duodenum
5. Intestinal juice digestive action in small intestine

＊ **Question 5.4**

(a) What is a fat? **(5 marks)**

(b) What are the functions of fats in living organisms? **(5 marks)**

(c) Describe (i) the digestion, and (ii) the absorption of *fats* in mammals. **(15 marks)**

(L)

Answer to Q5.4

(a) *Fats* are solid forms of lipids at room temperature, and are organic compounds of carbon, hydrogen and oxygen present in two main components, fatty alkanoic acids and glycerol (propanetriol).
Lipid fat = fatty alkanoic acid + glycerol (propanetriol)

(b) The functions of fats are:
1. *energy storage* material in plant seeds and fruits or under skin and around mammal kidneys – adipose tissue;
2. *cell membranes* are composed of lipids combined with proteins.

(c) (i) Fat digestion.

Mouth: Fat melts.

Stomach: Some lipase activity in *children*.

$$\text{Lipids} \xrightarrow[\text{pH 1 (acid)}]{\text{lipase}} \begin{array}{c}\text{alkanoic (fatty) acids}\\ +\\ \text{glycerol (propanetriol)}\end{array}$$

Small intestine:

$$\text{Lipid} \xrightarrow{\text{bile salts}} \text{emulsification, increasing surface area of lipid droplets for digestion}$$

$$\text{Lipid} \xrightarrow[\text{lipase}]{\text{pancreatic juice}} \begin{array}{c}\text{alkanoic (fatty) acids}\\ +\\ \text{glycerol (propanetriol)}\end{array}$$

(ii) *Absorption* of digestive products of fats in the form of tiny micelles, by diffusion, active transport and pinocytosis. To enter the central lacteal vessel which contains lymph. A diagram similar to the villus structure in Fig. 5.9 is needed here.

∗ **Question 5.5**

(a) Give an example of a food which is rich in protein and show how you would demonstrate this fact by means of a test. **(4 marks)**

(b) Describe the parts played by the following in the digestion and absorption of *protein* in the diet of a mammal: (i) the stomach; (ii) the pancreas; (iii) the small intestine. **(9 marks)**

(c) How do plants obtain their protein? **(2 marks)**

(SUJB)

Answer to Q5.5

(a) A food rich in protein is egg yolk (15%) or egg white (10%). Protein is *tested* for by mixing the egg with Biuret test reagents, when a violet colour appears.

(b) Protein digestion and absorption are summarised in tabular form. Diagram of villus (Fig. 5.9) also needed here.

Region of gut	*Digestion*	*Absorption*
Stomach	$\text{Protein} \xrightarrow[\text{gastric juice}]{\text{pepsins}} \text{peptides}$ (acid, pH 1) $\text{Milk} \xrightarrow[\text{(children)}]{\text{rennin}} \begin{array}{c}\text{casein}\\ \text{curds}\end{array}$ caseinogen	Amino acids can be absorbed in the stomach
Pancreas	$\text{Peptones} \xrightarrow[\text{pH 7}]{\text{trypsin}} \begin{array}{c}\text{amino acids}\\ +\\ \text{peptides}\end{array}$	None
Small intestine	$\text{Peptides} \xrightarrow[\text{pH 7–8}]{\text{peptidases}} \text{amino acids}$	Absorption by villi blood capillaries

(c) Plants can obtain their protein mainly by protein synthesis in the stem and root growing regions. Glucose undergoes a chemical change with nitrogen from soil nitrates and ammonium compounds to form amino acids. Some plants trap and feed upon insects (protein), i.e. *insectivorous* plants such as the sundew.

5.4 Self-Test Answers to Objective and Structured Questions

Answers to Multiple-choice Objective Questions

1. c   2. c   3. d   4. c   5. c   6. b   7. a   8. a   9. c   10. a   11. b   12. c
13. b   14. d   15. a   16. b   17. d   18. c   19. d   20. c

Answer to Structured Question 5.1

(a) D   (d) A
(b) A   (e) A
(c) C   (f) B

Answer to Structured Question 5.2

(a) (i)   Diagram A shows the *human* digestive system.
     (ii) Part L = colon; function is for cellulose digestion by gut micro-organisms. Part M = small intestine; function is mainly digestion and absorption.
     (iii) Diet of herbivore mainly plant material composed of *cellulose,* broken down into the monosaccharide *glucose* by action of micro-organisms.
     (iv) Pancreas produces powerful digestive enzymes: lipase, trypsin, amylase and nucleases.

(b) (i)   Tooth labelled Y is a *canine.*
     (ii) Total number of teeth shown in dental formula is 42.

$$\frac{3}{3} + \frac{1}{1} + \frac{4}{4} + \frac{2}{3} = 21 \times 2 = 42$$

6 Transport

6.1 Theoretical Work Summary

(a) Transport

Transport is the movement, circulation or flow of different substances, within living organisms, either with or without a continuous tubular vascular system.

(b) Transported Substances

1. *Radiant energy* into autotrophic green plants.
2. *Solutions* in water, of ions and small molecules of constructional, functional and waste materials.
3. *Solids* composed of large molecules of proteins and polysaccharides, *insoluble* suspensions, in blood cells, food particles, faeces, starch granules and chloroplasts.
4. *Gases* or vapours, oxygen, carbon dioxide and water vapour dissolved in water, or mixtures of gases in air.

(c) Cell Transport Methods

Cell membranes are *porous* living barriers that control the traffic into and out of cells which occurs by the mechanisms listed in Table 6.1.

Table 6.1   Mechanisms of movement of substances into and out of cells

Input/output method	Mechanism
DIFFUSION	Physical, passive — no energy needed, rapid in gases, slow in solutions
OSMOSIS	Physical, passive, slow for *water* only
ACTIVE TRANSPORT	Biological, energy-consuming process, *carriers* needed, molecules move rapidly from *low* concentration to *high* concentration region
FILTRATION	Physical, passive due to hydrostatic pressure, in kidney and blood vessels
PORES	Pores in cell membrane surface opening into *endoplasmic reticulum* of eukaryotes, pores in blood capillary walls and organelles

(d) Transport within Living Organisms

Transport of energy and chemical materials within living organisms is of two main kinds (see Table 6.2):

(a) *intracellular* transport '*within*' the cells of *all* organisms;

(b) *intercellular* transport '*between*' the cells and organs of *multicellular* organisms.

Table 6.2 Transport of energy and chemical materials in living organisms

Intracellular transport within cells	Intercellular transport between cells
Small amounts of small ions or molecules and water over short distances	Large amounts of large molecules, particles and water over long distances — *mass flow*
High surface area : cell volume ratio	Low surface area : body volume ratio
Vacuoles fluid filled, and contain particles in animals	Vessels fluid or blood filled
Diffusion in vacuoles or cytoplasm of ions and small molecules	Diffusion in *intercellular spaces*
Endoplasmic reticulum tubules transport large protein and lipid molecules. *Plasmodesmata* are narrow tubes connecting adjacent plant cells	*Vessels* transport large molecules in phloem of plants and vessels of animals; small molecules in plant xylem. *Pits* connect adjacent vessels in vascular plants
Cyclosis (circulation of cytoplasm) — food vacuoles in *Paramoecium*, and cytoplasmic streaming of chloroplasts in plants, e.g. in *Elodea* leaves	*Circulatory system* (heart and blood vessels in mammals). *Transpiration stream* and *translocation* in land plants
Microtubules associated with cell motion, and *microfilaments* in amoeboid movement	*Ciliated epithelium* transports mucus and small particles, e.g. ova, dust. *Peristalsis* occurs in tubular organs in the gut, uterus and ureters

(e) Transport in Flowering Land Plants

It is essential to know the *structure* of a flowering plant stem, root and leaf as described in Chapter 3 and section 4.1(e) before studying this topic. *Mesophytes* grow where water supply is adequate, neither too wet nor too dry; most flowering plants are mesophytes.

Table 6.3 summarises the main methods of transport for *intake*, *conduction* and *output* of energy and chemical substances.

Physical Processes

1. *Capillarity* is the *rising* of water in narrow, hair-diameter tubes, *xylem vessels*, due to the attractive force of *adhesion* between different substances, e.g. water and xylem lignin.
2. *Root pressure* results from osmosis of water into large numbers of root hairs on plant roots in spring when starch changes to glucose. Measured in kilopascals by *manometer* gauge.
3. *Leaf suction* or *transpiration pull* is a *force* due to water evaporating from leaf mesophyll cells, using heat energy; in turn it pulls a continuous unbroken

Table 6.3   Transport in flowering land plants (mesophytes)

Intake (see Fig. 6.1)	'Translocation'	Output
ABSORPTION or uptake of energy and chemical substances: ENERGY: solar radiant energy *refracted* by leaf epidermis into mesophyll cell chloroplasts CHEMICAL MATERIALS: Carbon dioxide, oxygen: diffusion from air via stomata into leaf mesophyll Water: soil water entry by *osmosis*, causing cell volume increase and *turgor* IONS: entry by *active transport* (therefore root hairs need oxygen), and by *nitrogen fixation* by root bacteria, or by *insect-trapping* in insectivorous plants *Pesticides* and *herbicides* enter through the plant surface	TRANSLOCATION, the *mass flow* or movement and circulation of soluble substances within multi-cellular *vascular* plants: 1. *Cell to cell* – osmosis of water, and diffusion of ions 2. *Intercellular spaces* and *porous cell walls* – capillary movement of water and solutes 3. *Xylem*: rapid, continuous, upward movement of water and solutes in transpiration stream (see Fig. 6.2) due to *leaf suction* 4. *Leaf mesophyll chloroplasts* receive water and ions and carbon dioxide for photosynthesis (*source* of organic material) 5. *Phloem* transports sucrose, amino acids, hormones and some ions to 'sinks' – growth and storage regions 6. *Pesticide* and *herbicide* chemicals are translocated by the xylem and phloem to kill either plant pests or the plant itself	HEAT ENERGY loss by radiation TRANSPIRATION, the loss of water from a plant surface: 1. *Osmosis* of water through turgid mesophyll cell walls. *Diffusion* of oxygen and carbon dioxide 2. *Evaporation* of water from mesophyll surface into leaf air space 3. *Diffusion* of water, carbon dioxide and oxygen from leaf air space into air via stomata, also through epidermis, cuticle and lenticels *Rate of transpiration* affected by: (i) temperature (ii) relative humidity (iii) wind speed (iv) light intensity (v) number of stomata (vi) soil water

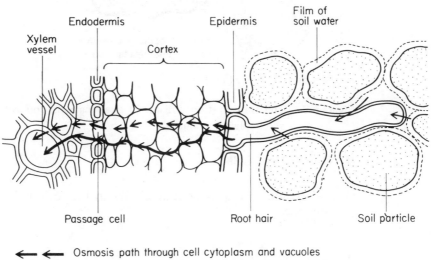

← ← Osmosis path through cell cytoplasm and vacuoles
←——— Diffusion path through cell wall pores and intercellular air spaces

Figure 6.1   The pathway of water from the soil into and across a flowering plant root to the xylem. In examinations you will probably be asked to 'label and explain this figure'.

column of water up the xylem vessels. Column held together by *cohesive* force of attraction between water molecules. Measured in kilopascals with cut shoot over mercury (Figs 6.2 and 6.10).

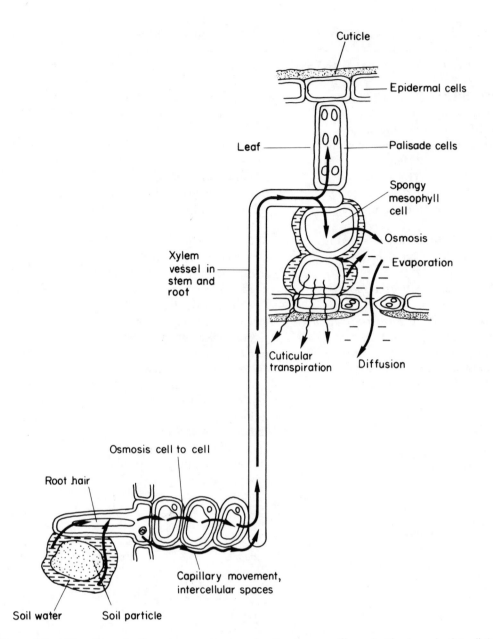

Figure 6.2 The transpiration stream. In examinations you will probably be asked to 'label and explain this figure'.

(f) Osmosis and Cells

Osmosis is illustrated diagrammatically in Fig. 6.3 and its effects are summarised in Table 6.4.

(g) Transport in Mammals

1. *Tissue fluid* (also called *intercellular fluid*) circulates in the spaces between animal body cells (Fig. 6.4).
2. *Intestine epithelial* cells take in nutrients from the gut cavity by various methods (listed in section 6.1(c), which are then expelled into the tissue fluid.

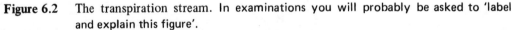

		INTESTINE	
GUT ⟶	CELL ⟶	EPITHELIAL ⟶	TISSUE
CAVITY	NUTRIENTS	CELL	FLUID

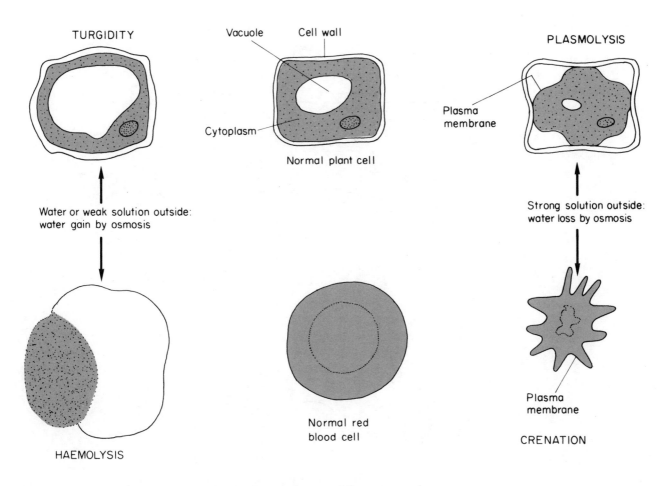

Figure 6.3 Osmosis in plant and animal cells.

Table 6.4 Effects of osmosis in plant and animal cells

Plant cells	*Animal cells*
(a) *Turgor*. Entry of water into vacuole by osmosis, when cell is in water, causing an increase in cell protoplasm volume and increased turgor pressure opposed by equal cell wall pressure	(a) *Haemolysis*. Entry of water into a red blood cell by osmosis, causing an increase in cell volume and bursting of plasma membrane, with escape of haemoglobin
(b) *Plasmolysis*. Loss of water from vacuole by osmosis, when cell is in a concentrated solution, causing a decrease in cell protoplasm volume and withdrawal of cell membrane from cell wall	(b) *Crenation*. Loss of water from a red blood cell when it is placed in a concentrated solution, causing a decrease in cell volume and shrinkage of plasma membrane

3. *Blood capillaries* in gut *villi* are in direct contact with the tissue fluid, which enters the capillaries rapidly through tiny pores, to pass on to *venules* leading to the heart.

4. *Lacteal* vessels collect a small amount of tissue fluid into the *lymphatic vessels*, which drains as *lymph* into the *vena cava*.

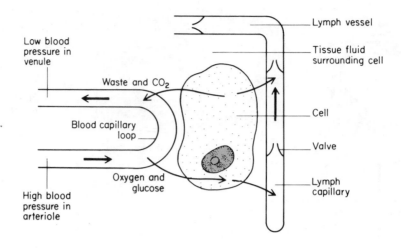

Figure 6.4   The formation of tissue fluid.

5. *Circulatory system* circulates:
(a) blood (Fig. 6.5 and Table 6.5);

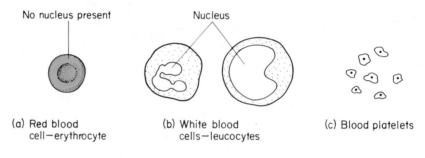

Figure 6.5   Mammalian blood cells.

Table 6.5 Human blood composition

Component Structure	Function
RED BLOOD CELL: *five million* cells in 1 mm³ blood. Formed in red bone marrow. Live 100 days.	Transport of oxygen as oxyhaemoglobin and carbon dioxide
WHITE BLOOD CELL: *seven thousand* cells in 1 mm³ blood. Several kinds. Formed in red bone marrow	Antibody formation; phagocytosis bacteria
PLATELETS: cell fragments	Release *fibrin* from *fibrinogen* in blood clotting
BLOOD PLASMA: forms 55% of blood, is 92% water and 7% protein	Transports nutrients, waste products, hormones and heat
BLOOD SERUM: plasma minus fibrinogen, a fluid separating from clotted blood	Used in the preparation of sera for vaccination
TISSUE FLUID: between living cells and similar to plasma but contains less protein	Cell *homeostasis* and transport
LYMPH: inside lymph vessels without protein, rich in lipids, contains white blood cells called *lymphocytes*	Transport to veins

(b) by *closed vascular system* (Fig. 6.6): blood being kept inside blood vessels, namely *arteries, arterioles, capillaries, venules* and *veins* (see also Table 6.6 and Fig. 6.7);

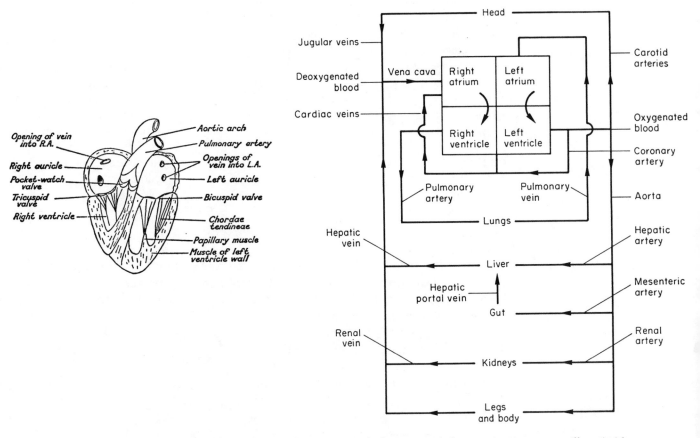

Figure 6.6 The heart and circulatory system of a mammal. In examinations you will probably be asked to 'label and explain this figure'.

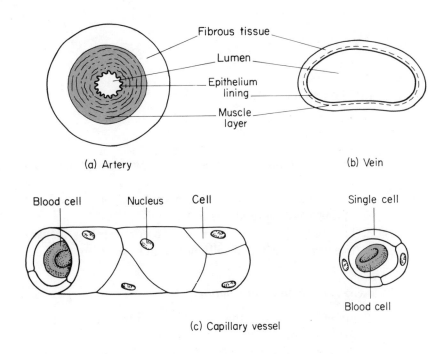

Figure 6.7 Internal structure of the three types of blood vessels.

Table 6.6 Comparison of arteries, veins and capillaries

Arteries	Veins	Capillaries
Thick, muscular walls	Thinner, less muscular walls	Walls of capillary one cell thick; no muscle
Elastic tissue	Less elastic tissue than in arteries	No elastic tissue
No valves (except in pulmonary arteries)	Valves present	No valves
Circular cross-section	Oval cross-section	Circular cross-section
Can contract	Cannot contract	Cannot contract
Not permeable	Not permeable	Permeable to tissue fluid and white blood cells
Transport blood *from* heart	Transport blood *to* heart	Transport blood from artery to vein
High pressure, 15 kPa	Low pressure, 2 kPa	Intermediate pressure, 5 kPa
Rapid flow	Slow flow	Slow flow
Pulse strong	No pulse	No pulse

(c) at *high speed under hydrostatic pressure* generated by *heart*, a pump composed of *two atria* (auricles) and *two ventricles* with *valves*;

(d) through a *double circulation*: the *systemic* circulation to *body*, and the *pulmonary* circulation to *lungs*:

(e) *heart* has a blood supply from the right and left *coronary arteries*; a *great cardiac vein* lies alongside the *left* coronary artery;

(f) *systole*: contraction stage of heart cycle; *diastole*: relaxation and dilation stage in heart cycle.

Note. AIDS disease is transmitted through contact with infected blood.

6.2 Practical Work

(a) Transport in Plants

(i) *Leaf Water Loss*

1. *Dry* blue cobalt chloride or thiocyanate paper turns *pink* in the presence of water.
2. The cobalt chloride or thiocyanate paper is fixed to upper and lower epidermis of a potted plant leaf as in Fig. 6.8.
3. The time taken for the paper to turn pink is measured for both leaf surfaces. The lower surface usually acts quickest on the test paper.

(ii) *Transpiration Channel*

1. Stand a white deadnettle plant with its roots, or a celery stem, in eosin solution for 2 hours.
2. Cut transverse sections of the stem as described in section 3.2(a)1 and examine with a microscope to show location of red eosin in the xylem vessels.

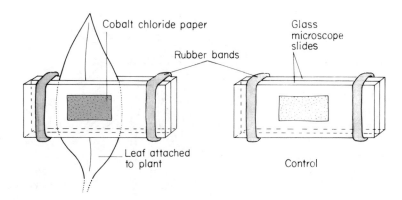

Figure 6.8 Demonstration of water loss from a dicotyledon leaf.

(iii) *Root Pressure Demonstration*

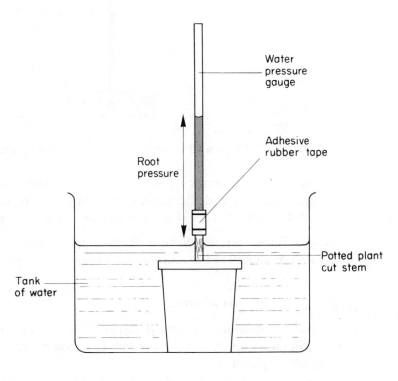

Figure 6.9 Apparatus to demonstrate root pressure.

The demonstration of root pressure (see Fig. 6.9) is performed in early spring.

1. Potted vine or fuchsia plant is submerged in a trough of water and the stem is cut clean across with a sharp knife or secateurs.
2. A tall, glass, water-filled tube or glass manometer is connected by adhesive rubber tape and firmly tied with wire or screw clips.
3. The water level rises in the glass tube or manometer, indicating increasing root pressure or hydrostatic pressure.

(iv) *Leaf Suction Force or Transpiration Pull*

1. A cut shoot of *Rhododendron* is connected to a narrow glass tube containing cooled and boiled water to expel dissolved air. The open end is placed in a dish of mercury (Fig. 6.10).

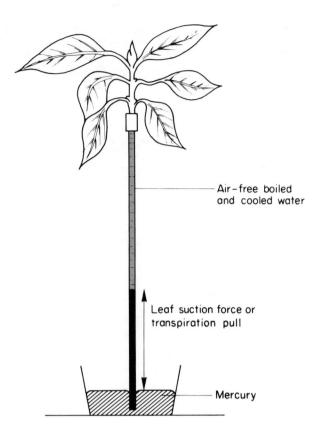

Figure 6.10　Apparatus to demonstrate transpiration pull.

2. As water evaporates from leaf mesophyll cells, the water column is pulled up the glass tube and the mercury rises after it. The height of the mercury column is a measure of the leaf suction force in centimetres of mercury.

Note: Since mercury vapour is poisonous, this experiment should be demonstrated by qualified instructors who observe strict precautions.

(v) *Water Loss of Transpiration*

1. The transpiration apparatus shown in Fig. 6.11 (and also a similar apparatus in structured question 6.1) is used to measure the weight of water lost from a plant or shoot.

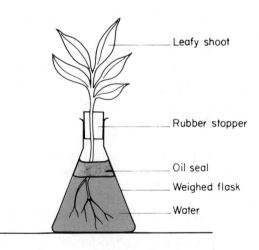

Figure 6.11　Mass or weight potometer apparatus to find out how much water a plant shoot loses by transpiration.

2. Variable *external conditions* can be arranged by keeping the apparatus in (a) a bright sunny position, (b) a dark cupboard, (c) close to a hairdryer set to 'cold', (d) close to a hairdryer set to 'warm'.

(vi) *Water Input Rate*

1. *Potometers* are used to determine the *rate of water input* or absorption by a cut shoot; it is normally closely related to the rate of water lost by transpiration (Fig. 6.12).

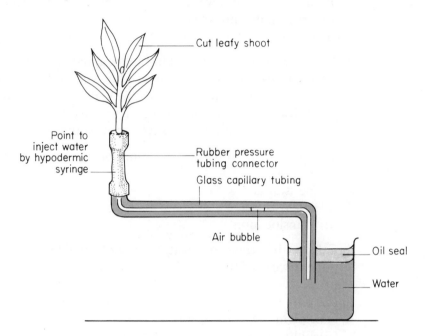

Figure 6.12   A water potometer for measuring rate of water uptake by a cut shoot.

2. Rate of water input is measured by the movement of the air bubble along the capillary tube.
3. External conditions can be varied as in (v).

(b) Transport in Animals

(i) *Blood Vessels*

Capillaries with moving blood in them can be seen under low power of a microscope in the thin skin close to the finger-nail base. A drop of cedarwood or almond oil added to the skin makes the capillaries easier to see.

(ii) *Blood Cells*

CARE! This experiment should only be done in accordance with local education authority regulations. Prepared microscope slides or photomicrographs should be examined as safer alternatives.
1. Use an edge of a microscope slide to smear a drop of blood on another clean microscope slide.
2. Leave to dry quickly in air, 2 min, add two drops Leishman's stain and leave for 5 min. Wash off excess in water. Cover with a cover slip and view through a microscope.

6.3 Examination Work

(a) Multiple-choice Objective Questions

1. Water loss in plants is most rapid when conditions are:
   - (a) wet, windy and cold
   - (b) wet, windy and warm
   - (c) dry, windy and warm
   - (d) dry, still and warm

2. The function of tissue fluid in the mammalian body is:
   - (a) to control the blood volume by leaking out of the blood vessels
   - (b) to carry food and oxygen to the cells from the blood
   - (c) to leak out of the blood and appear as sweat to cool the skin
   - (d) to surround and moisten cells in the body

3. Transport of food particles in *Paramoecium* occurs by means of:
   - (a) osmosis
   - (b) peristalsis
   - (c) cyclosis
   - (d) translocation

4. Cells which transport water upwards to the leaf from the roots are component parts of the:
   - (a) cortex
   - (b) pith
   - (c) xylem
   - (d) phloem

5. The movement of water from a film on the mesophyll cell surface into the intercellular air spaces of a leaf is a process called:
   - (a) osmosis
   - (b) diffusion
   - (c) evaporation
   - (d) pinocytosis

6. Which one of the following is a characteristic that applies to arteries?
   - (a) thin walls
   - (b) contain valves
   - (c) end as capillaries
   - (d) blood in them under low pressure

7. Which of the following is a characteristic that applies to mammalian red blood cells?
   - (a) nucleus present
   - (b) thin disc shape
   - (c) no haemoglobin present
   - (c) phagocytic action

8. One of the following is responsible for transport of sugars from the leaves to the roots of a flowering plant:
   - (a) xylem
   - (b) cambium
   - (c) phloem
   - (d) pith

9. The part of a root hair which acts as a semipermeable membrane is the:
   (a) cell wall
   (b) cell vacuole
   (c) cell membrane
   (d) cell nucleus

10. The name of the process by which water enters the root hair is called:
    (a) pinocytosis
    (b) active transport
    (c) osmosis
    (d) phagocytosis

11. The main advantage of a double circulation of blood in mammals is that:
    (a) there are no capillaries; tissues are bathed in blood
    (b) all parts of the heart share the work
    (c) oxygenated and deoxygenated bloods are mixed together thoroughly
    (d) the blood returns rapidly and more frequently to the heart

12. Which one of the following gases or vapours does *not* normally pass out of a leaf through the stomata?
    (a) carbon dioxide
    (b) oxygen
    (c) water
    (d) ammonia

13. Carbon dioxide is transported in the mammal blood plasma in the form of:
    (a) carboxyhaemoglobin
    (b) oxyhaemoglobin
    (c) hydrogencarbonates
    (d) sodium carbonate

14. The fluid component of the blood is called:
    (a) tissue fluid
    (b) serum
    (c) lymph
    (d) plasma

15. Red blood cells are made in a mammal in the:
    (a) liver
    (b) spleen
    (c) yellow bone marrow
    (d) red bone marrow

16. The blood pressure is *greatest* in the:
    (a) pulmonary artery
    (b) aorta
    (c) inferior vena cava
    (d) jugular vein

17. Transpiration is *least* affected by:
    (a) light intensity
    (b) wind speed
    (c) air temperature
    (d) air oxygen concentration

18. The component of blood that carries oxygen is the:
    (a) plasma fluid
    (b) white blood corpuscles

(c) red blood corpuscles

(d) platelets

19. The major blood vessel that takes deoxygenated blood to the heart from the body is called the:

(a) aorta

(b) vena cava

(c) coronary artery

(d) pulmonary vein

20. The concentration of a certain mineral ion in soil water is 50 parts per million (ppm). Inside the root hair cell vacuole, the same nutrient ion concentration is 200 ppm. This nutrient enters the root hair from the soil water by:

(a) diffusion

(b) active transport

(c) osmosis

(d) filtration

(b) Structured Questions

∗ **Question 6.1**

The apparatus shown is used in an experiment to investigate water relationships in plants.

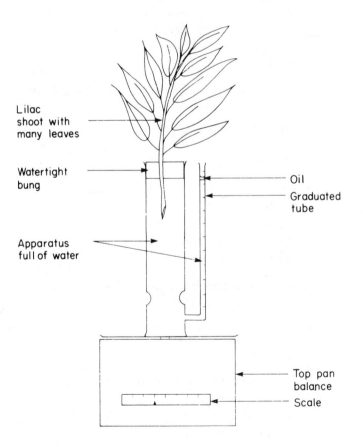

(a) What is the function of the oil in the graduated tube?

(b) What is the function of the balance?

(c) What would happen in the graduated side arm if water were taken up by the plant?

(d) Name the process which this apparatus is designed to measure.

(e) Give an explanation of what would happen to the balance reading if the apparatus were set up in normal laboratory conditions for 24 hours.

(f) Give an explanation of what would happen to the balance reading if the apparatus were moved to a warm, windy place for 3 hours.

(g) A polythene bag of negligible mass was placed over the aerial parts of the shoot and securely fastened to the neck of the water container. How would this affect the balance reading?

(h) Describe a suitable control experiment.

(i) Describe an experiment to demonstrate which side of the leaves has the greater water loss. **(14 marks)**

(NISEC)

Question 6.2

Stages in transpiration from mesophyll cells to the air outside the leaf are shown by arrows A, B and C in the accompanying diagram of a section through part of a leaf.

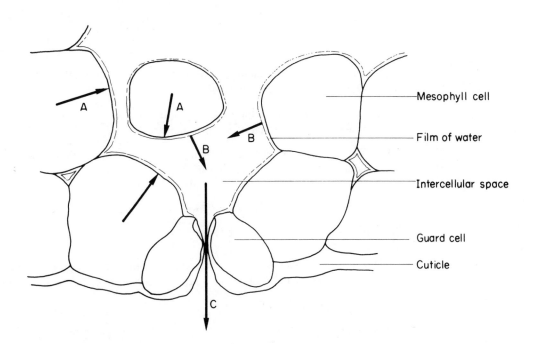

(a) What term best describes:
A, the movement of water from the mesophyll-cell vacuole to a film on the cell surface?
B, the movement of water from a film on the cell surface to vapour in the intercellular space?
C, the movement of water vapour from the intercellular space to the air outside?

(b) Briefly describe one function of the cuticle of the leaf. **(4 marks)**

(AEB, 1981)

✴ Question 6.3

(a) Name each of the parts of the mammalian heart labelled A–G in the accompanying diagram. **(7 marks)**

(b) What is the function of C? **(2 marks)**

(c) Why is the ventricular wall thicker than the auricular (atrial) wall? **(2 marks)**

(d) Why is the left ventricular wall thicker than the right ventricular wall? **(2 marks)**

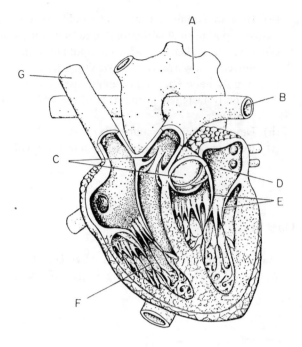

(e) Describe an experiment you would perform to investigate how the rate of heart-beat changes as the rate of work done by a mammal changes. **(4 marks)**

(f) Explain fully the value to the mammal of increasing the rate of heart-beat as the rate at which the mammal does work increases. **(4 marks)**

(g) What factors, other than changing the rate of work, influence the rate of heart-beat, and what is the effect of each factor? **(4 marks)**

(OLE)

✳ Question 6.4

The diagram shows the relation between a blood capillary, cells of the body and the lymph system in a mammal. Thick arrows show the direction of blood and lymph flow.

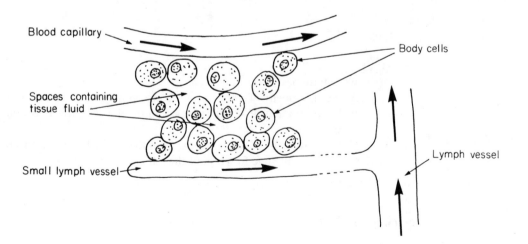

(a) Describe fully the structure of a blood capillary. **(5 marks)**

(b) How is the flow of blood maintained in a blood capillary? **(3 marks)**

(c) The table that follows shows the composition of blood. What is the main difference in composition between blood plasma and tissue fluid? How is the difference brought about? **(2 marks)**

	45% solid	Erythrocytes (red blood corpuscles)
		Leucocytes (white blood corpuscles)
		Thrombocytes (platelets)
Blood		
	55% fluid plasma	47.0% water
		7.0% protein
		0.9% salt
		0.1% glucose

(d) Describe a leucocyte which moves from the blood capillary through tissue fluid into the lymph system. How is this movement brought about and how does this leucocyte leave the blood capillary? Describe the functions of this leucocyte when it is in the tissue fluid. **(10 marks)**

(e) What is lymph? What is the major difference between lymph and tissue fluid? Where does lymph return to the blood circulation? **(4 marks)**

(f) The larger veins and lymph vessels contain valves. Describe the structure of one of these valves and explain how, with the help of valves, blood and lymph are returned to the heart. **(6 marks)**
(AEB, 1981)

(c) Free-response-type Questions

(i) *Short-answer Questions*

1. (a) Why do trees usually die when a complete ring of bark is removed from around the trunk? **(3 marks)**

(b) Give one way in which this damage could happen naturally. **(1 mark)**
(OLE)

2. (a) Draw two views of a mammalian red blood corpuscle. Describe the views that you have drawn.
(b) Briefly describe the contents of a red blood corpuscle.
(c) What is the main function of a red blood corpuscle? **(5 marks)**
(AEB, 1981)

3. State *three* ways in which the transport system of a flowering plant differs from that of a mammal. **(6 marks)**
SUJB)

(ii) *Long-answer Essay Questions*

∗ **Question 6.5**

(a) State three ways in which water is important to flowering plants. **(3 marks)**

(b) What structural features may lead to a reduction of water loss from a leaf? **(8 marks)**

(c) Describe an experiment to measure the rate of water loss from a leafy shoot. **(11 marks)**

(d) What conditions would increase the rate of water loss from the leafy shoot used in this experiment? **(3 marks)**
(L)

Answer to Q6.5

(a) Water is important to flowering plants (i) for cell turgidity in plant support; (ii) as a raw material for photosynthesis; (iii) as a solvent for ions; (iv) for transport; and (v) for transpiration pull.

(b) Structural features which reduce water loss from a leaf are:
   (i)   thick epidermal cuticle
   (ii) small needle-like leaves
   (iii) rolled leaves
   (iv) hairs on leaf surface
   (v) stomata sunk into leaf epidermis

(c) Water loss from a leafy shoot can be measured by the transpiration apparatus (Fig. 6.11) or as in structured question 6.1.
   Note: Question does not ask for an experiment to show water *uptake* as in the potometer.

(d) Conditions increasing water loss rate: (i) low relative humidity of air, (ii) high wind speed, (iii) high air temperature.

✳ Question 6.6

Describe the structure and functions of the heart and blood vessels in a mammal. **(25 marks)**
(L)

Answer to Q6.6

The key words in this direct question are *structure*, *function*, *heart* and *blood vessels*.

Structure of Heart

The essential diagram to show the heart's structure is that given in the answer to structured question 6.3, showing the heart in *vertical* section.

Function of Heart

The heart is a muscular pumping organ for blood circulation.

Structure of Blood Vessels

This requires Fig. 6.7 showing artery, vein and capillary vessels in transverse section. Need to mention presence of *valves* in veins not shown in diagrams.

Function of Blood Vessels

Arteries transport blood from heart to tissues and usually carry oxygenated blood (exception *pulmonary* artery). *Veins* transport blood from tissues to heart and usually carry deoxygenated blood (exception *pulmonary* veins). *Capillaries* between arterioles and venules drain blood into venules and then into veins.

✱ Question 6.7

    (a) What is double circulation?   **(2 marks)**

    (b) Contrast the structure and function of an artery with that of a vein.   **(8 marks)**

    (c) It is estimated that the total length of the capillaries in man is 60 km. They are much branched, thin-walled and narrow. Explain the significance of these features.   **(5 marks)**

                                                                       (SUJB)

Answer to Q6.7

(a) *Double circulation* is the circulatory system of birds and mammals, composed of two parts:
   (i) pulmonary circulation to *lungs* from heart;
 (ii) systemic circulation to *body* from heart.

(b) Figure 6.7 needed here, and also the following:

Artery	Vein
Structure	
Thick, muscular, elastic walls; can contract. No valves	Thin, less elastic walls; cannot contract. Valves present
Function	
Blood transport from heart, at high pressure. Mainly oxygenated (exception pulmonary artery)	Blood transport to heart, at low pressure. Mainly deoxygenated (exception pulmonary vein)

(c) The estimated 60 km of capillary vessels will provide;
    (i) close contact through thin walls with almost all body cells through large surface area;
   (ii) rapid supply of nutrients and oxygen for tissue fluid formation by diffusion, osmosis and hydrostatic blood pressure;
  (iii) rapid collection and removal of cell waste from tissue fluid.

6.4 Self-test Answers to Objective and Structured Questions

Answers to Multiple-choice Objective Questions

1. c   2. d   3. c   4. c   5. c   6. c   7. b   8. c   9. c   10. c   11. d   12. d
13. c   14. d   15. d   16. b   17. d   18. c   19. b   20. a

Answers to Structured Question 6.1

(a) The graduated tube contains oil to prevent evaporation of water.

(b) The balance records the loss in weight of the transpiration apparatus.

(c) The water level in the graduated side arm falls as water is taken up by the shoot.

(d) This apparatus is designed to measure the weight of water loss by transpiration.

(e) The balance reading would indicate a loss in weight of the transpiration apparatus during the 24-hour period in normal conditions.

(f) When in a warm, windy place, the loss in weight would be very rapid compared with that in normal laboratory conditions.

(g) A polythene bag would enclose and hold the water evaporated from the shoot and there would be no significant weight loss shown by the balance.

(h) A suitable control would be the same transpiration apparatus fitted with a complete, watertight bung *without* a lilac shoot.

(i) Fix blue cobalt chloride papers to upper and lower surfaces of a leaf. The *lower* surface transpires most rapidly and will turn pink before the upper surface.

Answer to Structured Question 6.2

(a) A, osmosis; B, evaporation; C, diffusion.

(b) Cuticle functions are to prevent water loss through the epidermis, and entry of water into epidermis.

Answer to Structured Question 6.3

(a) A, aorta; B, pulmonary artery; C, semi-lunar valves; D, left atrium or auricle; E, bicuspid or mitral valve; F, chordae tendineae *or* the right ventricle (either accepted); G, vena cava.

(b) They are valves to allow blood to flow in one direction – no back flow.

(c) To provide muscle force on contraction.

(d) Left ventricle wall is thicker to provide greater force to drive blood to all parts of body; right ventricle forces blood a short distance to the lungs.

(e) Take pulse of person at rest, then take pulse after the person has done work (going up and down stairs three times).

(f) Increased heart-beat rate increases rate of *supply* and *removal* of tissue fluid to body cells; providing oxygen and glucose, and removing waste products: CO_2, H_2O and excretory products.

(g) *Fright or fear* increases heart-beat rate and prepares the body for action to fight or run away, under the influence of the hormone adrenalin. *Carbon dioxide* gas present in air: increasing concentrations cause an increase in heart-beat rate.

Answer to Structured Question 6.4

(a) Capillary blood vessels form from arterioles and are one cell in thickness.

(b) Blood flow maintained by pressure in arteries and arterioles.

(c) Tissue fluid contains more water (a content of nearly 98%) than blood plasma, due to the former having no solid blood cells. Also there is little if any protein since large molecules and blood cells are unable to pass through the capillary walls.

(d) Leucocyte moves by amoeboid movement by squeezing itself between the spaces where the capillary cell walls meet. The leucocytes function protectively as phagocytes in the blood.

(e) Lymph is the fluid inside the lymphatic vessels. It is similar to tissue fluid but contains lipids and white blood cells. It returns to the general blood circulation at a point in the vena cava.

(f) Valves are membranous structures allowing flow in one direction; there are semi-lunar valves in veins and lymph vessels. Blood and lymph return to the heart by the compression effect of surrounding skeletal muscles.

7 Respiration

7.1 Theoretical Work Summary

(a) Food Energy Release

Respiration is an enzyme-controlled process of breakdown (catabolism) of high-energy-content organic food (mainly glucose) occurring internally in tissues and cells of living organisms. The products of respiration are:

1. *ATP (adenosine triphosphate)*, a high-energy-value compound and source of cell energy.
2. *Heat energy* or warmth.
3. Different *respiratory products*: carbon dioxide, water, ethanol and lactic acid.

(b) Adenosine Triphosphate, ATP

One molecule of ATP will release 50 kJ of *free energy* for cell use, by enzymatic hydrolysis or interaction with water.

$$\text{ATP, adenosine } tri\text{phosphate} + \text{Water and enzyme} \longrightarrow \substack{50 \text{ kJ} \\ \text{ENERGY} \\ \text{for} \\ \text{cell use}} + \text{ADP, adenosine } di\text{phosphate} + \text{PHOSPHATE ION}$$

ATP is *re-formed* from ADP and phosphate ion using energy from sunlight or other sources.

$$\text{ADP} + \text{PHOSPHATE} + \substack{\text{ENERGY} \\ (50 \text{ kJ})} \longrightarrow \text{ATP} + \text{water}$$

(c) Types of Cell Respiration

Respiration in cells and tissues is of two main kinds (see Table 7.1):

(i) *Anaerobic Respiration*

Anaerobic respiration means food breakdown *without* oxygen. Two different processes are possible:

1. *Yeasts*, and for short periods higher green plant cells, produce ethanol and carbon dioxide from glucose:

$$\substack{\text{GLUCOSE} \\ C_6H_{12}O_6} \longrightarrow \substack{\text{ETHANOL} \\ 2C_2H_5OH} + \substack{\text{CARBON} \\ \text{DIOXIDE} \\ 2CO_2} + \text{ENERGY}$$

2. Bacteria in sour milk, and for short periods *skeletal muscle* of diving mammals such as whales and seals, and mudworms produce lactic acid from glucose.

GLUCOSE ⟶ LACTIC ACID + ENERGY

$C_6H_{12}O_6$ ⟶ $2CH_3.CHOH.COOH$

Table 7.1 A comparison of anaerobic and aerobic respiration

	Anaerobic respiration	*Aerobic respiration*
ORGANISM	Yeasts and, for short periods, green plants Bacteria and worms; diving mammals in skeletal muscle for short periods	All air-breathing living organisms. Yeasts can also respire aerobically
LOCATION	Cell cytoplasm	Cell mitochondria
ATP from 1 molecule of glucose	2 ATP = 100 kJ/mol	38 ATP = 1900 kJ/mol Nineteen times more energy
HEAT ENERGY	Forms 35–50% of total energy released	35% of total energy released
RESPIRATORY PRODUCTS	*Ethanol and carbon dioxide:* as in manufacture of beers, wines and spirits, and in flour-dough raising *Lactic acid*: toxic in large amounts. NO metabolic water formed	*Carbon dioxide and metabolic water:* 100 g lipid forms 107 g H_2O 100 g carbohydrate forms 55 g H_2O 100 g protein forms 41 g H_2O
OXYGEN	Not used	Essential

(ii) *Aerobic Respiration*

Aerobic respiration means food breakdown *using oxygen*. It occurs in cells of all air-breathing living organisms and produces carbon dioxide and water (called metabolic water) from glucose.

GLUCOSE + OXYGEN ⟶ CARBON DIOXIDE + WATER + ENERGY

$C_6H_{12}O_6$ + $6O_2$ ⟶ $6CO_2$ + $6H_2O$

(d) Energy-providing Nutrients

Other substances apart from glucose are sources of cell energy (see Table 5.2) by aerobic respiration. The energy value per gram of carbohydrate is 17 kJ, protein 17 kJ, lipids 38 kJ. The *products* of *anaerobic* respiration have the following energy values: ethanol 29 kJ/g, lactic and fruit acids 15 kJ/g.

(e) Oxygen Debt

Skeletal muscle produces lactic acid by anaerobic respiration when the tissue is temporarily short of oxygen. An excessive amount of lactic acid in muscle causes fatigue and muscle pain. Lactic acid requires oxygen to respire it aerobically in the liver; the intake of oxygen therefore continues for some time after severe exercise to repay the oxygen debt. (See Fig. 7.1.)

Table 7.2   Gas exchange in aquatic and air-breathing animals

Organism	*External Respiration*
Aquatic organisms UNICELLULAR ANIMALS	Water contains 1% dissolved oxygen Water flow supplies and removes oxygen; body movement brings this about in stagnant water. Diffusion of gases through *plasma membrane*; small flattened body with large surface area to volume ratio
FISH	Water flow, by mouth *pumping* and *sucking action*. Continuous water stream over large surface area of *vascularised gill filaments* and flaps on gill arches. Rapid diffusion into gill-filament capillary vessels (Fig. 7.2)
Respiration in air MOIST SKINNED	Air contains 20% oxygen Earthworms and frogs; diffusion of air gases through thin, mucus-moistened skin of large surface area, into blood capillary vessels. (Frog uses mouth lining and lung epithelia for gas exchange — mouth acts as air pump.)
INSECTS	Air drawn into tubular *tracheae* by way of *spiracles* to exchange with tissue by fluid-filled *tracheoles*. Air movement by diffusion or muscular contraction causing changes in volume of abdomen (Fig. 7.3)
BIRDS	Draw in air by body muscle movement into *air sacs*, to supply lungs which function for gas exchange in a way similar to that in mammals. No diaphragm — flight muscles used instead
MAMMALS	(i) *Ventilation* is by combined action of *diaphragm* and *intercostal muscles*, causing change in volume of thorax and rib cage, with changing internal air pressure, drawing in air (inspiration), expelling air (expiration) (Fig. 7.4) (ii) *Gas exchange* occurs across lung *alveoli* epithelium by diffusion gradients between air in blood capillaries and air in alveoli. Water evaporates into alveoli (Fig. 7.5)

(f) External Respiration

External respiration is the process by which oxygen is brought to the internal aerobic respiration centres or *mitochondria* of aerobic plants and animals. Two main processes may be involved.

1. *Ventilation*: transport of gases to and from cells by air, water or blood fluids.
2. *Gaseous exchange* (see Table 7.2): takes place at cell surface or respiratory organ surface.

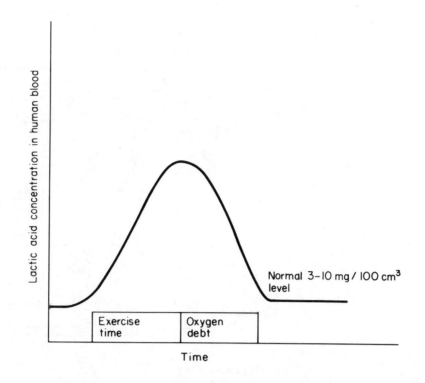

Figure 7.1 Graph to show the development of lactic acid concentration due to insufficient oxygen. This shortfall is repaid by increased intake during the oxygen debt period.

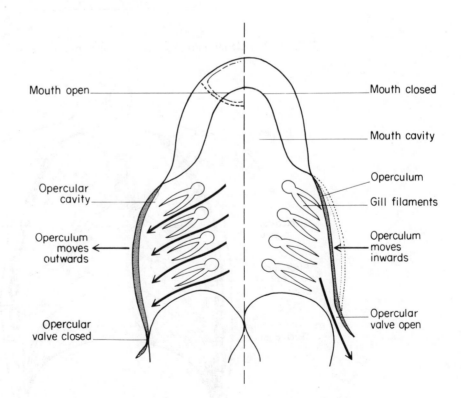

Figure 7.2 Ventilation movements in a bony fish.

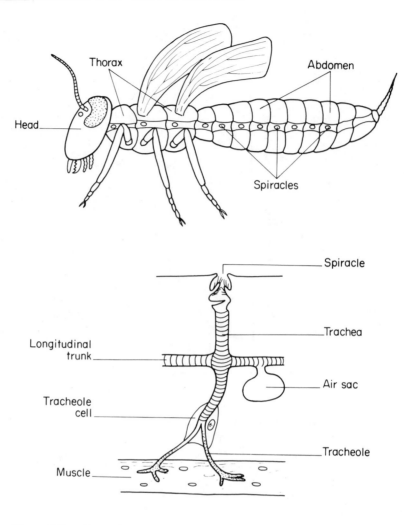

Figure 7.3   General arrangement of the respiratory system of an insect.

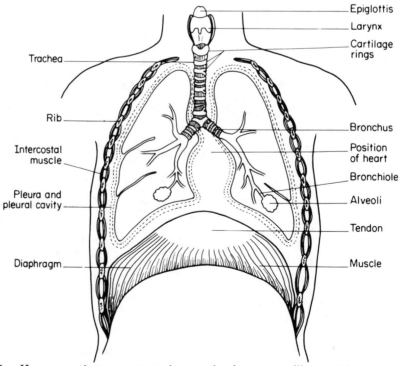

Figure 7.4   Human respiratory system. In examinations you will probably be asked to 'label and explain this figure'.

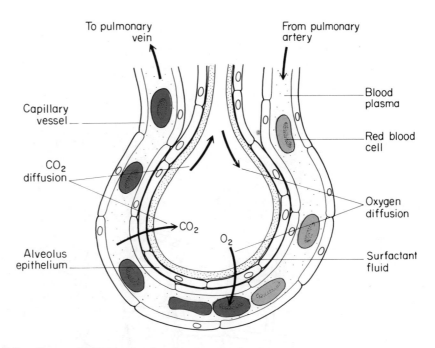

Figure 7.5 Gas exchange at the alveolus. In examinations you will probably be asked to 'label and explain this figure'.

Characteristics of Gas-exchange Surfaces

1. Large surface area of body or provided by respiratory *organs*.
2. Thin permeable unicellular or membranous layer.
3. Moist surface.
4. Blood or other fluid transport system nearby.
5. Gas diffusion gradient from high to low concentration.

Table 7.3 Summary of respiratory movements in man

Part of respiratory tract	*Inspiration*	*Expiration*
Diaphragm	Contracts and flattens downwards	Relaxes and moves upwards to dome shape
Intercostal muscles	External intercostal muscles contract	Internal intercostal muscles contract
Rib cage and sternum	Move upwards and outwards	Move downwards and inwards
Thorax volume	Increases	Decreases
Air pressure	*Decrease* in pressure inside thorax and lung	*Increase* in pressure inside thorax and lung
Air movement	External air pressure drives air into lungs at low pressure	Air forced out of lungs by thorax *compression* and *elastic recoil* of lungs
Pleural cavity	Pleural fluid lubricates pleural membranes	Pleural fluid lubricates pleural membranes

(h) Plant Gas-exchange Methods

Terrestrial Green Plants

1. *Oxygen* is a by-product of photosynthesis.
2. *Diffusion* of oxygen occurs over a short distance from *chloroplast* to *mitochondrion*, when plant is in *daylight*.

3. *Respiration* occurs when plant is in light or darkness. Air gases diffuse via *stomata* and *lenticels* (Fig. 7.6) to intercellular air spaces to reach living plant cells.

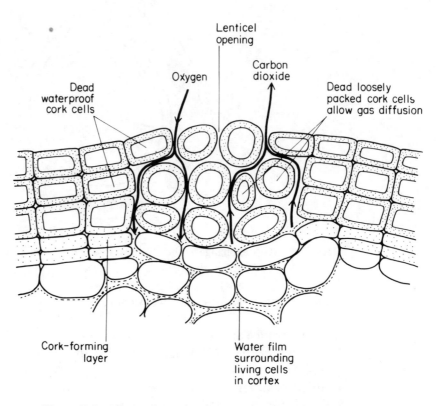

Figure 7.6   Gas exchange in the stem lenticel of a flowering plant.

4. *Root hairs* have large *surface area* to allow exchange of gases beween *soil air* and living cell.

Aquatic Green Plants

1. In submerged plants without stomata, gas exchange is by diffusion.
2. Floating leaves have stomata on *upper* epidermis.

Non-green – Fungi

1. Are *without oxygen* supply from chloroplasts since they do not photosynthesise.
2. Gas exchange is entirely by diffusion into air.

(i) Comparison of Respiration in Green Plants and Animals

In Table 7.4 we compare the process of respiration as it occurs in green plants and animals.

Table 7.4 Respiration in green plants and animals

Green-plant respiration	Animal respiration
No *external* ventilation movements	*External* ventilation movements in all except small animals, e.g. *Amoeba* and *Hydra*
Gaseous exchange is on cell surface of uni-cellular plants, and in air spaces between cells in leaf or stem and root hairs of multicellular plants	Gaseous exchange is on cell surface of uni-cellular animals, or in lungs, skin, gills or by air tubes in multicellular animals
No special transport system	Blood systems transport oxygen
Cellular respiration produces energy formed as ATP in mitochondria	Cellular respiration produces energy formed as ATP in mitochondria
Aerobic respiration forms carbon dioxide and water	Aerobic respiration forms carbon dioxide and water
Anaerobic respiration forms ethanol and carbon dioxide	Anaerobic respiration forms lactic acid (2-hydroxypropanoic acid) only, and *no* carbon dioxide
Green plants produce little detectable heat	Animals produce considerable detectable heat
Aerobic plants have additional oxygen source from photosynthesis	Aerobic animals have one oxygen source, the air
Respiration rate low	Respiration rate high

Energy Usage

1. DNA replication, cell division and growth (Chapter 8).
2. Active transport and maternal uptake (Chapter 6).
3. Protein synthesis from amino acids.
4. Heat for maintenance of body temperature in animals (Chapter 9).
5. Muscle contraction in animals (Chapter 11).
6. Nerve-impulse transmission in animals (Chapter 10).

(j) Respiration and Photosynthesis Compared in Green Flowering Plants

1. The maximum rate of photosynthesis in a green flowering plant is almost 30 times the rate of respiration. Consequently the increase in amount of organic material results in growth and an abundant *oxygen* supply (Fig. 7.7).
2. *Compensation point* is the *light intensity* at which the rate of carbon dioxide *uptake* (photosynthesis) is *exactly equal* to the rate of carbon dioxide *production* (respiration). At the compensation points in the early morning and at dusk, the net gain of carbohydrate from photosynthesis will be equal to the net loss of carbohydrate by respiration.
3. Table 7.5 compares the processes of respiration and photosynthesis.

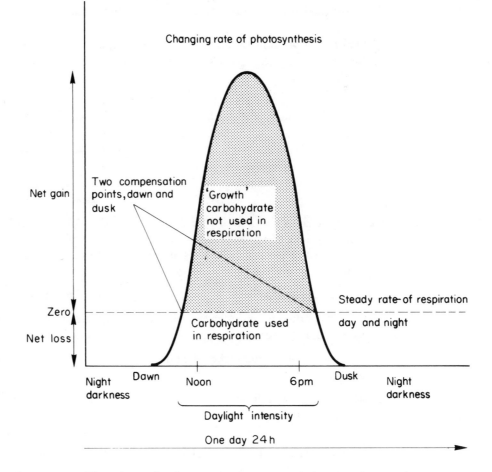

Figure 7.7 The relationship between respiration and photosynthesis in flowering plants.

Table 7.5 A comparison of photosynthesis and respiration

Photosynthesis	Respiration
A process of *anabolism* or synthesis of organic materials	A process of *catabolism* or breaking down of organic materials
Energy trapped from sunlight	Energy released and stored as ATP
Occurs in cells containing *chloroplasts*	Occurs in all cells containing *mitochondria*
Light essential	Occurs at all times in light or dark
Oxygen produced	Oxygen needed for *aerobic* respiration
Carbon dioxide required	Carbon dioxide produced by plant and animal aerobic respiration, and plant anaerobic respiration
Typical of all green plants and certain bacteria	Typical of all living organism aerobes and anaerobes
Variable rate with varying light intensity	Unaffected by changing light intensity

(k) Basal Metabolic Rate

(a) *Metabolic rate* is the rate at which a body uses energy, and can be measured as either oxygen uptake or carbon dioxide output, using a respirometer or spirometer.

(b) *Basal metabolic rate*, BMR, is the rate at which a body uses energy when at complete rest, 12 hours after a meal, in a moderately warm room. In such conditions energy is used for: heart beat, external respiration and cell metabolism. Basal metabolism excludes all physical skeletal muscle activity.

Factors Affecting BMR

1. Thyroid gland produces thyroxine, and the gland's over-activity (excess thyroxine) increases BMR.
2. Age — BMR highest in babies, decreases with age.
3. Sex — females have lower BMR.
4. Motherhood — BMR increases in pregnancy, and in lactating women.
5. Drugs: smoking and nervous stimulants increase the BMR.
6. Species of organisms. The BMR varies in the different species.
7. Small animals have a higher BMR than larger animals.

7.2 Practical Work

(a) Respirometer

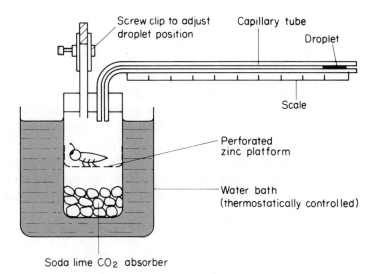

Figure 7.8   A simple respirometer.

1. The *simplest* apparatus (see Fig. 7.8) consists of a glass container with a narrow-bore capillary tube fitted in a rubber stopper.
2. *Carbon dioxide* is absorbed in soda lime granules beneath a perforated zinc platform supporting the specimen.
3. *Control* experiment is a similar apparatus either empty or containing glass beads equal in volume to live specimen.
4. Both experiment and control apparatus are maintained in a water bath at 37°C. Since the volume of oxygen taken up = volume of carbon dioxide produced in respiration, the *decrease* in volume is seen as movement of a liquid drop along

127

the capillary tube. The *time* taken for the droplet to move a *measured distance* is recorded.

Metabolic rate = rate of oxygen uptake

Increasing *temperatures* will increase the metabolic rate up to an *optimum* temperature.

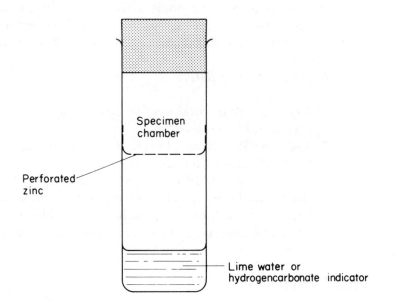

Figure 7.9 Apparatus to find out if carbon dioxide is produced by small living organisms.

(b) Air Composition

The air composition (see Table 7.6) can be tested using the following methods:
1. Carbon dioxide absorbed either by soda lime, sodium hydroxide or potassium hydroxide solutions. Also by barium hydroxide (baryta), and by clear calcium hydroxide (lime water) solution, which turns cloudy.
2. Oxygen *and* carbon dioxide absorbed together in alkaline pyrogallol.
3. Nitrogen and other inert gases (argon or neon) *not* soluble in any reagent.
4. Graduated glass gas-measuring *burettes* are used for air composition measurement (see Table 7.6).
   (a) Carbon dioxide content equals volume reduction over NaOH, KOH or soda lime.
   (b) Oxygen content equals volume reduction over alkaline pyrogallol *after* treatment in (a) to absorb carbon dioxide.
   (c) Nitrogen and inert gases are the remaining volume of unabsorbed gas.

Table 7.6 Gaseous composition of air

Gas	Inspired room air	Expired lung air	Air in lung alveolus
O_2	20%	15%	13%
CO_2	0.03%	4.0%	5.3%
N_2	78%	75%	75%

(c) Formation of Carbon Dioxide in Respiration

1. Green plants must have respirometer surrounded by light-proof metal foil to *prevent* photosynthesis.
2. *Specimen*: green leaf, fruit, flower, germinating seed or small animals, supported on perforated zinc platform in *respirometer* (Figs 7.8 and 7.9).
3. *Clear* calcium hydroxide or *red* hydrogen carbonate indicator replaces soda lime. Carbon dioxide causes lime water to turn milky, hydrogencarbonate indicator *yellow*.

(d) Heat Production in Respiration

1. *Live*, germinating and *dead* (boiled) peas are *sterilised* by rinsing in dilute hypochlorite solution to remove bacteria and mould fungi.
2. *Narrow-range* thermometer inserted among the pea seeds in a vacuum flask, *inverted* to prevent heat loss.
3. *Control* consists of boiled, sterile pea seeds.
4. Heat cannot enter vacuum flask: heat can only come from germinating seeds. (See part (c) of model answer to Q7.4.)

(e) Yeast Anaerobic Respiration

Experiment

1. Flask contains 3% yeast in a 10% glucose solution made from *sterile* water previously boiled and cooled, to *expel* dissolved oxygen. Oil seal prevents oxygen reaching yeast cells by diffusion (oxygen is *insoluble* in oil). (See Fig. 7.10.)

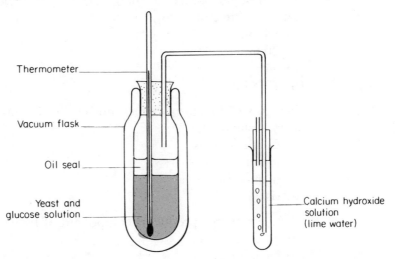

Figure 7.10   Apparatus to show anaerobic respiration in yeast, *Saccharomyces* sp.

2. *Control* is a boiled and cooled solution of yeast and glucose. Boiling kills yeast cells.
3. *Heat* production in anaerobic respiration detected by narrow-range thermometer; heat cannot enter vacuum flask from outside.
4. *Ethanol* can only be detected at this stage by its *smell*. (Sample can be collected by *distilling* filtered yeast mixture.) Sample burns with pale blue flame. (Lactic acid has *no* smell and has an acid taste).

7.3 Examination Work

(a) Multiple-choice Objective Questions

1. In comparison with inspired air, air leaving a mammalian alveolus contains:

	Carbon dioxide	Oxygen	Water vapour
(a)	less	less	more
(b)	less	more	less
(c)	more	less	less
(d)	more	less	more

2. When a mammal breathes out:
   (a) muscles in the bronchioles contract
   (b) the diaphragm becomes dome shaped
   (c) the epiglottis covers the oesophagus opening
   (d) the lung air volume increases

3. Oxygen is essential for the:
   (a) intake of water by root hairs
   (b) intake of ions by root hairs
   (c) transpiration of water from leaves
   (d) development of root pressure

4. A human being at complete rest has an energy turnover of 4 kJ per minute. A gram of glucose has an energy value of 16 kJ/g. How long will 1 g of glucose last as a source of energy?
   (a) 64 min
   (b) 20 min
   (c) 4 min
   (d) 12 min

5. The class of organisms which possess tracheoles as a component of the respiratory system are:
   (a) tracheophytes
   (b) insects
   (c) mammals
   (d) fish

6. The main energy-producing component of tissue fluid is:
   (a) glucose
   (b) an amino acid
   (c) a neutral lipid
   (d) cholesterol

7. Which of the following is a product of anaerobic respiration in mammal skeletal muscle?
   (a) glucose
   (b) lactic acid
   (c) ethanol
   (d) ethanoic acid

8. A green plant at night in darkness has a leaf input of one of the following from air:
   (a) nitrogen
   (b) carbon dioxide
   (c) oxygen
   (d) water vapour

9. When the human body is engaged in very strenuous physical exercise, the blood sugar glucose is broken down into energy and:
   (a) carbon dioxide and water
   (b) lactic acid
   (c) glycogen
   (d) ethanol and carbon dioxide

10. During the first few days of growth of a germinating broadbean seedling it is seen to show a decrease in dry weight. This is due to:
    (a) photosynthesis and organic material formation
    (b) transpiration from the first-formed foliage leaves
    (c) respiration of food reserves
    (d) formation of new tissues

11. Energy for direct use in living cells is available as:
    (a) adenosine diphosphate, ADP
    (b) adenosine triphosphate, ATP
    (c) glucose
    (d) glycogen

12. The organelle of eukaryote cells associated with aerobic respiration is called the:
    (a) ribosome
    (b) lysosome
    (c) mitochondrion
    (d) chloroplast

13. Soil air is essential for all flowering plants in order to:
    (a) provide nitrogen for protein formation
    (b) provide carbon dioxide for photosynthesis
    (c) provide oxygen for root hair and root respiration
    (d) receive ethanol vapour from anaerobic roots

14. A method for determining the metabolic rate of an animal is to measure one of the following over a period of time:
    (a) water output
    (b) oxygen intake
    (c) food intake
    (d) body mass decrease

15. The minimum amount of energy on which the human body can survive is called:
    (a) kinetic energy
    (b) potential energy
    (c) basal metabolic rate
    (d) food energy value

16. Which of the following is the correct balanced chemical equation for cell aerobic respiration?

    (a) $6CO_2 + 6H_2O \longrightarrow C_6H_{12}O_6 + 6O_2$

(b) $6CO_2 + 6O_2 \longrightarrow C_6H_{12}O_6 + 6H_2O$

(c) $6O_2 + C_6H_{12}O_6 \longrightarrow 6H_2O + 6CO_2$

(d) $O_2 + C_6H_{12}O_6 \longrightarrow 6H_2O + CO_2$

17. One gram of pure lipid produces the following amount of energy by aerobic respiration in the human body:

(a) 16 kJ/g

(b) 17 kJ/g

(c) 28 kJ/g

(d) 38 kJ/g

18. The approximate percentage of the energy content of glucose which is made into ATP by aerobic respiration is:

(a) 20%

(b) 50%

(c) 75%

(d) 100%

19. The water that forms as a product of aerobic respiration is called:

(a) soil water

(b) transpired water

(c) metabolic water

(d) tissue fluid

20. The main source of oxygen in the cells of Canadian waterweed submerged in brightly illuminated stagnant pond water *deficient* in oxygen is the cell:

(a) mitochondrion

(b) chloroplast

(c) vacuole

(d) nucleus

(b) Structured Questions

Question 7.1

The respiration of living organisms can be studied using apparatus such as that shown in the diagram.

(i) What chemical could be placed at A to absorb all carbon dioxide from the air entering the apparatus?

(ii) If active blowfly larvae were placed in the experimental chamber and the pump was run for several hours, what change would be seen in the apparatus?

(iii) If you were to use this apparatus to investigate the respiration of a potted plant, what precautions would you take?

(iv) Explain the need for these precautions. **(6 marks)**

(SUJB)

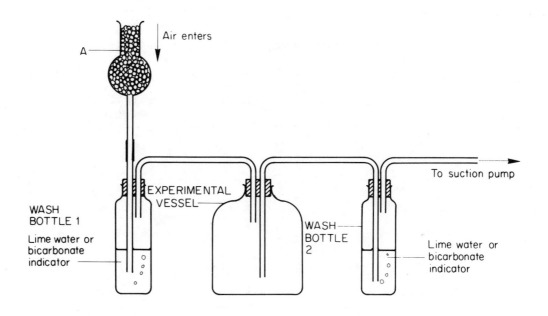

Question 7.2

The diagram shows the apparatus used in an investigation into anaerobic respiration in yeast. The suspension was made up in water that had been boiled to ensure that no oxygen was present at the start of the experiment. The liquid paraffin prevents oxygen from entering the suspension during the experiment.

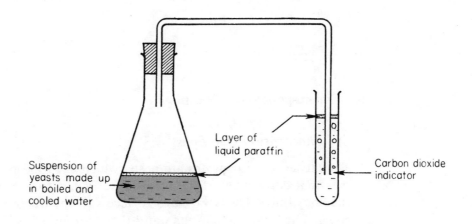

(a) Why can carbon dioxide pass from the suspension through the liquid paraffin when oxygen cannot enter the suspension from the air in the flask?
(b) Name a substance which could be used as the carbon dioxide indicator and describe the change you would observe in this substance if carbon dioxide were produced.
(c) What other substance should be included in the suspension to ensure the production of carbon dioxide? **(4 marks)**

(AEB, 1983)

* Question 7.3

The table shows data collected from an athlete after a period of running.

Running time (min)	Concentration of lactic acid in blood (mg/100 cm³)
0	3
2	12
4	29
6	46
7	50
8	52
9	50
10	45
12	41
14	37
18	31
23	22
28	18
35	13
45	9
55	5
65	3

(a) Make a suitable graph of the data. **(8 marks)**

(b) What process and which tissue are responsible for the production of lactic acid in the body? **(2 marks)**

(c) Comment on the events occurring in the body between the 8th and 65th minutes of the investigation. **(5 marks)**
(SUJB)

(c) Free-response-type Questions

(i) Short-answer Questions

1. The larvae of flour moths are pests, which eat food products such as porridge oats that contain very little water, insufficient to explain the amount found in the body fluids. The larvae have no access to any other external source of water. Explain how they obtain the water they require to meet their needs. **(2 marks)**
(SUJB)

2. (a) What are the percentages by volume of oxygen, argon, carbon dioxide and nitrogen in the atmosphere? **(4 marks)**

(b) Which of these gases is most soluble in water? **(1 mark)**

(c) How does a temperature rise affect the solubility of a gas? **(1 mark)**

(d) How does dissolved carbon dioxide affect the pH of water? **(2 marks)**
(OLE)

3. Give four pieces of evidence that would lead you to believe that a particular organ in an animal was used for exchange of respiratory gases. **(4 marks)**
(OLE)

4. Most animals have specialised structures which act as gas-exchange surfaces.
   (a) Name the structure at which gas exchange occurs in: (i) a mammal; (ii) an insect.
   (b) State two features these structures have in common. **(4 marks)**
(AEB, 1983)

(ii) *Long-answer Questions*

✻ **Question 7.4**

(a) Give a *balanced* equation which summarises the process of tissue respiration. **(5 marks)**

(b) What are the differences between aerobic and anaerobic respiration? **(7 marks)**

(c) Describe an experiment which demonstrates that germinating seeds give off heat. **(9 marks)**

(d) How does a named protozoan obtain its oxygen supply? **(4 marks)**
(L)

Answer to Q7.4

(a) It is important to note the *key* words in this question, 'balanced equation'. It asks for a *chemical* equation:

$$C_6H_{12}O_6 + 6O_2 \longrightarrow 6H_2O + 6CO_2 + ENERGY$$

It is a *balanced* equation because there are the same number of atoms of different chemical elements each side of the $\longrightarrow$ sign.

$6 \times C \longrightarrow 6 \times C$

$12 \times H \longrightarrow 12 \times H$

$18 \times O \longrightarrow 18 \times O$

REACTANTS $\longrightarrow$ PRODUCTS

(b)

Aerobic	Anaerobic
1. Most organism cells	1. Certain bacteria, fungi and skeletal muscle
2. Products are carbon dioxide and water	2. Products are ethanol and carbon dioxide or lactic acid
3. High energy production	3. Low energy production
4. Oxygen needed	4. Oxygen not needed

135

(c) The apparatus shown in the accompanying diagram is needed here.

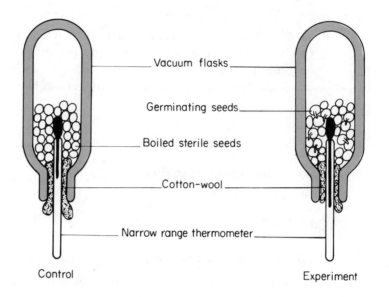

(d) A *named* protozoan could be *Amoeba*, which obtains its oxygen supply from the surrounding water which has an oxygen concentration of 1% approximately. Entry of oxygen is by way of the ectoplasm or *plasma membrane* by *diffusion*.

✳ **Question 7.5**

Respiration and photosynthesis are vital chemical processes occurring in plant cells.

(a) Show in tabular form the differences between these processes. **(5 marks)**

(b) Describe how you would show experimentally that carbon dioxide is evolved as a result of aerobic respiration in a green potted plant. Mention any precautions and controls you would use. **(10 marks)**
(SUJB)

Answer to Q7.5

This *mixed* question draws on knowledge of plant nutrition (Chapter 4) and respiration.
(a)

Respiration	*Photosynthesis*
Mitochondria	Chloroplasts
Energy release	Energy trapped
Catabolism	Anabolism
Light not essential	Light essential
Oxygen needed	Oxygen produced

Five differences are listed to obtain five marks awarded in this question section.

(b) The diagram of the apparatus which *must* be drawn is that shown in structured question 7.1, since the potted plant is large and could not fit into the small respirometer shown in Fig. 7.9.

Essential Precautions

1. Keep potted plant in *total darkness* by wrapping experimental chamber in light-proof material.
2. Wrap pot with gas-proof metal foil up to plant stem, to prevent escape of carbon dioxide from pot soil.
3. *Control* performed in separate apparatus using a wrapped pot of soil *only* i.e. without a plant.

✽ Question 7.6

(a) Make a labelled diagram to show the anatomy of the mammalian thorax. **(17 marks)**

(b) Explain fully how a mammal breathes. **(8 marks)**
(OLE)

Answer to Q7.6

(a) A popular direct question requiring a large labelled diagram showing the internal structure of a mammalian thorax. This means that all structures composing it inside and out must be indicated: Fig. 7.4 is needed here.
  (i) *Thorax* — rib, intercostal muscle (external and internal), diaphragm, thoracic cavity.
  (ii) *Lungs* — nostril, nasal cavity, epiglottis, larynx, trachea, bronchi, bronchioles and alveoli. Pleural membranes. Indicate position of oesophagus.

(b) The word 'breathes' means ventilation, and a description of breathing *movements* in inspiration and expiration is needed:

	Inspiration	*Expiration*
Intercostal muscle	Externals contract	Internals contract
Ribs	Raised	Lowered
Diaphragm	Contracts or flattens	Relaxes or domes
Internal thorax pressure	Decreases	Increases
Thorax volume	Increases	Decreases
Air flow	Inwards	Outwards

7.4 Self-test Answers to Objective and Structured Questions

Answers to Multiple-choice Objective Questions

1. d　2. b　3. b　4. c　5. b　6. a　7. b　8. c　9. b　10. c　11. b　12. c
13. c　14. b　15. c　16. c　17. d　18. b · 19. c　20. b

Answer to Structured Question 7.1

This diagram shows the apparatus used to demonstrate the evolution of CO_2 during respiration of a *large* organism which could not be put into the smaller respirometer apparatus of Fig. 7.9.

(i) The chemical placed at A to absorb carbon dioxide entering the apparatus is *soda lime*.

(ii) Changes seen after several hours with respiring blowfly larvae in experimental vessel – clear lime water or red bicarbonate indicator in wash bottle 1, cloudy limewater or yellow bicarbonate indicator in wash bottle 2.

(iii) Potted plants must have the pot completely wrapped in gas-proof material up to the plant stem, and the experimental vessel completely blacked out with light-proof material such as aluminium foil.

(iv) The reason for wrapping the pot is to prevent escape of carbon dioxide produced by soil organisms in respiration. The plant is prevented from carrying out photosynthesis by excluding light and keeping it in the dark.

Answer to Structured Question 7.2

(a) Carbon dioxide passes through the liquid paraffin oil layer as an insoluble gas under pressure which is greater than the surrounding air pressure due to its increasing volume as it is formed. Oxygen is unable to pass by diffusion since a gas must be soluble in oil before diffusion can occur, and oxygen is insoluble in oil. The air pressure is too low to affect the solubility of oxygen in oil.

(b) Substances: clear lime water or red hydrogen carbonate indicator. Changes: lime water turns cloudy, and red indicator yellow.

(c) The yeasts must have a respiratory *substrate*, therefore glucose should be added to the suspension.

Answer to Structured Question 7.3

A graph is constructed from the data in the table, with time on the horizontal axis and lactic acid concentration on the vertical axis.

(a) The graph *form* is similar to that in Fig. 7.1.

(b) Anaerobic respiration is responsible for lactic acid formation in the skeletal muscle tissue.

(c) Between the 8th and 65th minutes, there will be muscle fatigue, cramp or pain and need to pant to repay the oxygen debt in order to change lactic acid into glucose using some of the energy obtained by aerobic respiration of part of the lactic acid.

8 Growth

8.1 Theoretical Work Summary

(a) Growth

Growth is the irreversible increase in bulk, dry weight and complexity of an organism. Its processes and effects are summarised in Table 8.1.

Table 8.1 Growth processes and their effects

Growth processes	Effect
1. Synthesis	Formation of *structural* organic materials, plant cellulose, animal structural proteins, by photosynthesis and protein synthesis *Storage deposits* of starches, lipids and minerals, e.g. calcium in bone
2. Cell division	*Mitosis*: division of cell nucleus and cytoplasm with duplication of nuclear material
3. Cell enlargement	Plant cells form vacuoles, increasing cell size and volume
4. Cell differentiation	Growth of unspecialised cells into special cells in *tissues*

Growth results when:
(i) *anabolism* (the synthesis of organic substances) is greater than *catabolism* (the breakdown of organic substances by respiration);
(ii) *inputs* of energy and raw materials are greater than *outputs* of energy and secretory and excretory products.

(b) Mitosis

Mitosis is the equal division of the cell nucleus and cytoplasm into two separate cells, after a cell has reached an optimum size. It is important in the growth and *asexual* reproduction of organisms (Fig. 8.1).

Interphase is the period in cell life when the cell nucleus material, DNA (deoxyribose nucleic acid) is *doubled* in quantity, i.e. when the *chromosomes* are said to *replicate*.

Mitosis occurs as a continuous process over four main phases, *prophase, metaphase, anaphase* and *telophase*. In plant cells a new *cell wall* forms (as in Fig. 8.2) along a middle lamella.

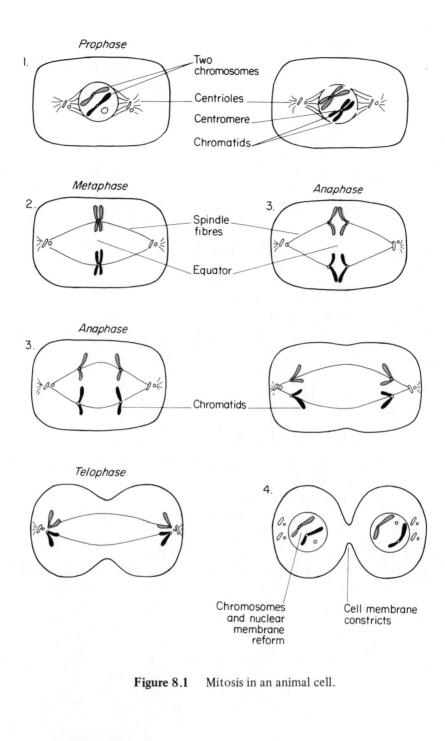

Figure 8.1 Mitosis in an animal cell.

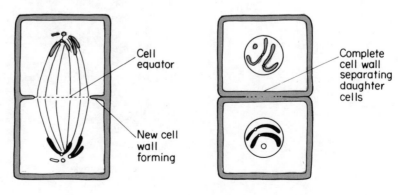

Figure 8.2 Cell wall formation after mitosis in a plant cell.

The cell *life cycle* can be summarised as follows:

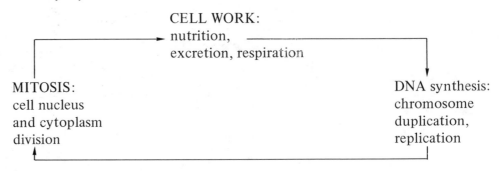

CELL WORK:
nutrition,
excretion, respiration

MITOSIS:
cell nucleus
and cytoplasm
division

DNA synthesis:
chromosome
duplication,
replication

(c) Growth in Flowering Plants

Growth in flowering plants takes place by means of three processes:
1. *germination*, i.e. seed growth;
2. *primary growth* from primary *apical meristems* of root and shoot tips;
3. *secondary growth* from secondary *lateral meristems* of vascular cambium.

(d) Seed Growth

(i) Seed Structure

The structure of some typical flowering plant (angiosperm) seeds is shown in Figs 8.3 and 8.4.
1. The *embryo* consists of:
   (a) *plumule*, the future shoot
   (b) *radicle*, the future root
   (c) *cotyledon(s)*, one or two seed leaves
   (d) *epicotyl*, which connects plumule to cotyledon stalk
   (e) *hypocotyl*, which connects radicle to cotyledon stalk

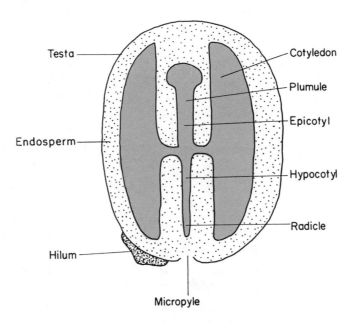

Figure 8.3 Generalised structure of a seed.

142

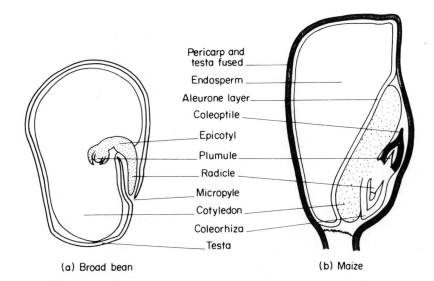

Pericarp and testa fused
Endosperm
Aleurone layer
Coleoptile
Epicotyl
Plumule
Radicle
Micropyle
Cotyledon
Coleorhiza
Testa

(a) Broad bean   (b) Maize

Figure 8.4   Internal structure of (a) a dicotyledon seed, broad bean, *Vicia faba*, and (b) the monocotyledon fruit (grain) of maize, *Zea* sp. In examinations you will probably be asked to 'label and explain these figures'.

2. The *endosperm* is the nutritive tissue to provide energy and nutrients for the growth of the developing embryo. In the *endospermic seeds* of cereals and oil-seed plants, the endosperm consists of starch and oils absorbed by cotyledons during germination. *Non-endospermic* seeds (e.g. peas and beans) have food stored in cotyledons.
3. The *testa* (seed coat) has:
   (a) a *micropyle* for water entry;
   (b) a *hilum*, which connects seed to fruit wall.

(ii)   *Seed Germination*

The process of seed germination is illustrated in Fig. 8.5. In order for germination to occur, the following are necessary:
1. *water* to cause cell *enlargement* or vacuolation, for *hydrolysis* and *transport*;
2. *warmth* at optimum temperature for enzyme activity;
3. *oxygen* from soil air for energy release (aerobic respiration);
4. *red light* needed by some seeds, e.g. lettuce.
5. *energy* and *nutrient supply* from food reserve.
   There are two types of germination (see Fig. 8.5):
1. *epigeal*, as in French and runner beans, where the cotyledons form the first photosynthetic leaves and the hypocotyl grows rapidly;
2. *hypogeal*, as in maize and legume seeds, where the cotyledons remain underground as food storage organs and the epicotyl grows rapidly.

(e) Primary Growth

Growth in *length* arising from *meristems* in root and stem tips or *apices* is described as primary growth. The *meristem* is a region of actively dividing cells. There are three main regions of primary growth:
1. *cell division* region in root and stem apex cells by mitosis;
2. *cell elongation* region where divided cells *vacuolate* with water intake (Fig. 8.6);
3. *cell differentiation* region where vacuolated cells become different *tissues* of cortex, epidermis and vascular tissue, seen in unthickened root and stem (see Fig. 3.2).

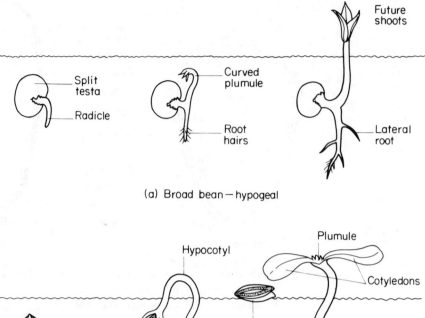

(a) Broad bean — hypogeal

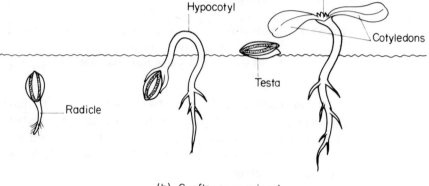

(b) Sunflower — epigeal

Figure 8.5   Stages in germination in seeds of (a) broad bean, *Vicia faba*, and (b) sunflower, *Helianthus* sp.

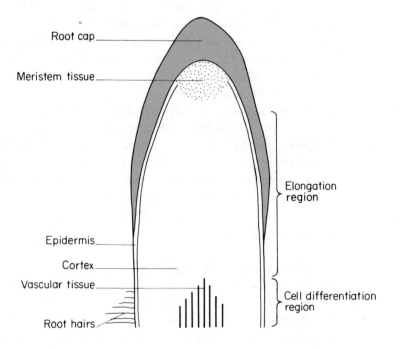

Figure 8.6   Longitudinal section through a root to show regions of cell division and elongation.

(f) Secondary Growth

Secondary growth is from the cells in the secondary meristems or *cambium* in the vascular tissue, which divide by *mitosis* resulting in *secondary thickening* or increase in girth of woody stems and roots (Fig. 8.7). It is seen as *growth rings* in many plants in temperate regions.

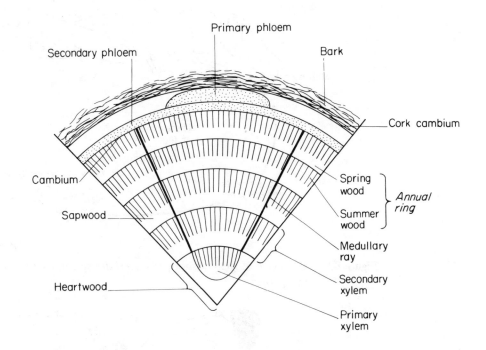

Figure 8.7   Transverse section of a portion of a woody stem showing secondary thickening.

(g) Growth in Multicellular Animals

There are three phases in the growth of a multicellular animal:
1. *cleavage* of a fertilised egg by mitosis to form a blastula, a ball of cells;
2. *gastrulation* of undifferentiated cells into ectoderm, endoderm and mesoderm, the germ layers;
3. *morphogenesis*, the development of organs, form and structure.

Metamorphosis

This is a stage in the animal life cycle between the egg and adult.

Incomplete metamorphosis takes place in insects such as cockroaches, locusts, grasshoppers and termites, in which there is a *gradual* change from *nymph* to adult *imago* (Fig. 8.8).

Complete metamorphosis occurs in the frog, and such insects as butterfly, housefly and moth. There is a *rapid* change from a *larva*, which is very different from the adult. Insects show a *pupa* stage from which the adult *imago* emerges (see Figs 8.9 and 8.10).

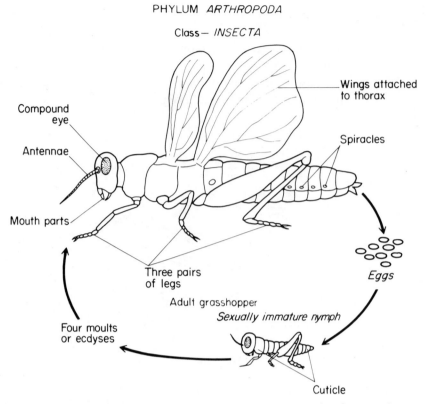

Figure 8.8   Life cycle of the grasshopper — *Acridiid* sp., an example of incomplete metamorphosis.

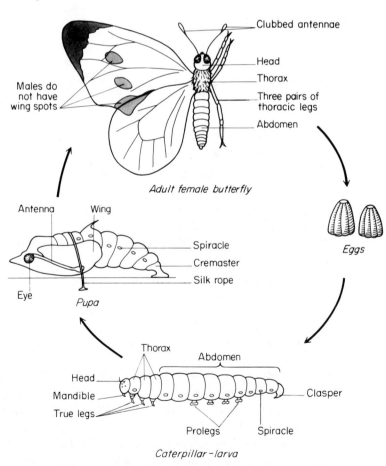

Figure 8.9   Life cycle of the large white butterfly, *Pieris brassicae*, which shows complete metamorphosis.

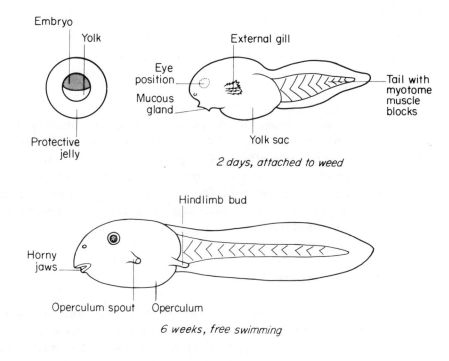

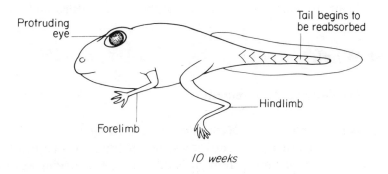

Figure 8.10 Outline of the life cycle of the common frog, *Rana temporaria*.

(h) Growth Measurement

Growth can be measured by the following methods:

1. Height or length by *auxanometer*.
2. Dry or fresh weight increase.
3. Volume increase.
4. Time lapse photography.

The *rate of growth* can be shown graphically by plotting growth increments over periods of time.

Determinate growth is seen in organisms which *stop* growing when a certain size is reached, as in annuals, birds and mammals. *Indeterminate growth* is *unceasing* growth (in trees and other perennial plants, fish and reptiles). *Intermittent growth* occurs in arthropods, which periodically shed the exoskeleton by *ecdysis*. *Allometric growth* is the growth of different body organs at different rates. Leaves grow more quickly than stems. Brain growth in human beings stops at age 5, but other body parts continue to grow. Reproductive organs have a rapid growth rate between 14 and 18 years.

(i) Growth Factors

The factors that affect the rate of growth are:
1. *Heredity* or inheritance of a trait (character) for tallness or dwarfness – see Section 13.1(e).
2. *Nutrition* – see Chapters 4 and 5.
3. *Light* is essential for plants in order for photosynthesis to occur; mammals show rickets (bone softness leading to stunted growth).
4. *Temperature* affects plant growth, but has no effect on animal growth.
5. *Hormones*. Plant hormones or growth substances called *auxins* cause *vacuolation* and *tropisms* (growth movement). Animal growth is affected by the growth hormones *thyroxine*, from the thyroid gland, and *somatotrophin*, from the pituitary gland.

(j) Dormancy

Dormancy is a state in which growth ceases and metabolism is at its lowest rate. Most plant buds, seeds, bulbs, spores, corms and tubers undergo dormancy. *Hibernation* is a type of *winter* dormancy that many mammals undergo. *Aestivation* is a type of dormancy that occurs during heat or drought in many desert-living animals, and in fish, amphibia, and reptiles such as crocodiles and alligators.

8.2 Practical Work

(a) Growth Measurement

An *auxanometer* records the increase in *length* of a plant part. The auxanometer lever magnifies growth over a scale, or traces it on a rotating drum.

Dry weight measurement is determined by drying samples at 110°C to constant weight. It is a measure of the organic substance produced.

A growth graph (e.g. Fig. 8.11) has a curve that is S-shaped and has four phases:

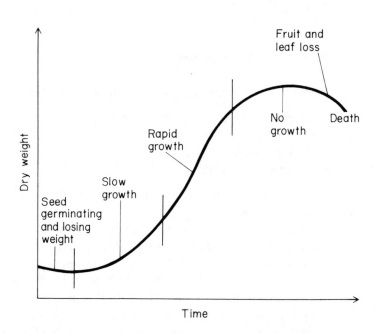

Figure 8.11 Graph to show stages in growth of an annual flowering plant.

(i)   slow growth
(ii) rapid growth
(iii) zero growth
(iv) senescence or degradation, preceding death

(b) Plant Growth

1. *Mitosis* and cell division are observed in prepared slides of root and stem apical meristems of broad bean.
2. The *elongation region* can be observed by marking the radicle of a germinating broad bean seedling (Fig. 8.12).

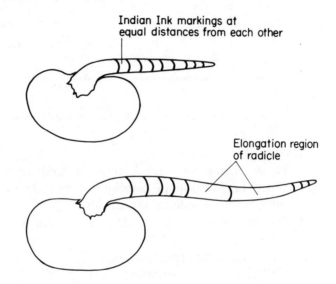

Figure 8.12   Determination of the growth region in a broad bean seed radicle.

3. *Tropic growth movement*
    (a) *Phototropism* is shown by oat or grass seedlings in response to one-sided illumination (Fig. 8.13).

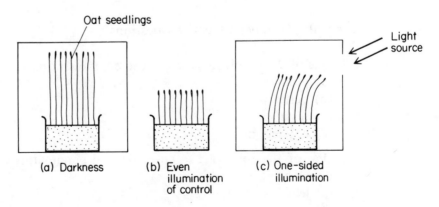

Figure 8.13   Apparatus to demonstrate phototropism in oat, *Avena* sp.

    (b) *Geotropism*, growth in response to the effect of gravity, is removed when a bean seedling is grown on a *klinostat. Slow* rotation of the seedling results in all its parts receiving an identical stimulus.

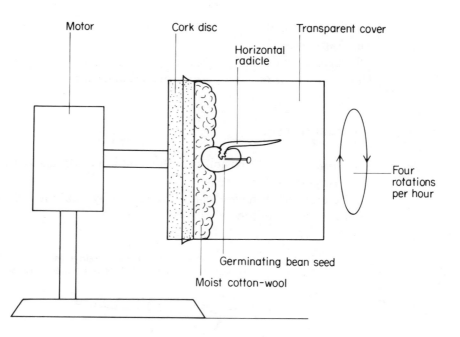

Figure 8.14 Experimental work with a klinostat to demonstrate geotropism in broad bean.

4. *Growth substances*

These substances, also known as auxins (plant hormones), when applied as dilute preparation in lanolin on one side of maize seedings, cause curvature away from the side of application.

5. The presence of *seed food reserves* can be tested using soaked seeds and the following:

   (a) dilute iodine for starches;

   (b) 'Albustix' for proteins;

   (c) Sudan III stain, or grease mark on paper for lipids.

6. *Secondary growth* is observed in prepared microscope slides of lime, *Tilia* sp., stems.

8.3 Examination Work

(a) Multiple-choice Objective Questions

1. One of the following is *not* a seed component part:
   (a) plumule
   (b) pericarp
   (c) testa
   (d) cotyledon

2. One of the following is *not* a natural perennating organ in a perennial deciduous plant:
   (a) leaf
   (b) bulb
   (c) seed
   (d) twig

3. The seed of a potato is found in the:
   (a) tuber
   (b) fruit

(c) stem

(d) leaf

4. The best definition of growth among the following is:
   (a) an irreversible increase in size or dry weight
   (b) an increase in food content
   (c) an increase in height and girth
   (d) cell size increase by water uptake

5. Metamorphosis is a stage when:
   (a) there are larval forms in a life cycle
   (b) gradual changes occur in organism development
   (c) there is a rapid change from adult to juvenile form
   (d) the body becomes segmented

6. During insect metamorphosis most growth is seen in the:
   (a) egg
   (b) larva
   (c) imago
   (d) pupa

7. A substance which causes different rates of growth on opposite sides of a plant stem is called:
   (a) an auxin
   (b) an enzyme
   (c) ATP (adenosine triphosphate)
   (d) an antibody

8. Seedlings which have been germinated in the dark and *kept* in the dark would soon show:
   (a) phototropism
   (b) etiolation
   (c) anaerobic respiration
   (d) photosynthesis

9. The hilum and micropyle are component structures of a:
   (a) fruit wall
   (b) seed coat
   (c) leaf blade
   (d) tuber surface

10. The hilum is the point of attachment of a:
    (a) fruit to a floral axis
    (b) leaf to a stem
    (c) tuber to a stem
    (d) seed to a placenta

11. The instrument designed to measure increase in length of a plant part is called:
    (a) an auxanometer
    (b) a potometer
    (c) a respirometer
    (d) a klinostat

12. Which of the following glands produces the growth hormone in mammals?
    (a) sweat
    (b) salivary
    (c) pituitary
    (d) thymus

13. One condition which is *not* essential for seed germination is:
    (a) suitable temperature
    (b) oxygen
    (c) water
    (d) carbon dioxide

14. Mitosis is the process of equal division of the cell:
    (a) wall
    (b) chloroplasts
    (c) nucleus
    (d) vacuole

15. Mitosis is necessary in the unicellular organism for:
    (a) tissue formation
    (b) growth in size
    (c) reproduction
    (d) vacuolation

16. In which of the following would mitosis be seen to be taking place:
    (a) human red blood cells
    (b) pollen grain cells
    (c) root-tip meristem cells
    (d) leaf mesophyll palisade cells

17. The process of periodic shedding of the rigid exoskeleton to allow growth is seen only in:
    (a) arthropods
    (b) birds
    (c) mammals
    (d) reptiles

18. During the process of cell growth, the nuclear or chromosome material called DNA doubles in quantity by the process of:
    (a) protein synthesis
    (b) replication
    (c) mitosis
    (d) vacuolation

19. Which of the following is a response to a stimulus occurring in vertebrate animals?
    (a) tropisms
    (b) taxisms
    (c) reflex action
    (d) diffusion

20. Which of the following would *not* show the process of mitosis in the cell nucleus?
    (a) stem meristem cells
    (b) seed endosperm cells
    (c) *Spirogyra* cell
    (d) bacteria cell

(b) Structured Questions

✳ **Question 8.1**

Three groups of seeds of identical weight were treated as follows:

Group X. Heated in an oven at 105°C until no further weight loss was noted. Final dry weight 6.5 g

Group Y. Soaked in water and left in light for 3 days. Small roots had appeared by this time. Then treated as group X. Final weight 5.9 g.

Group Z. Soaked in water and left in light for 8 days. Long roots and a green shoot had formed on each seedling at the end of this time. Then treated as group X. Final dry weight 7.1 g

Suggest what may have caused the difference between:

(i) the dry weights of X and Y.

(ii) the dry weights of Y and Z.

(iii) Apart from those mentioned, state TWO other factors which seeds require in order to germinate.

(iv) Distinguish between hypogeal and epigeal germination. **(8 marks)**

(SUJB)

Question 8.2

The accompanying diagrams represent a plant root tip and three cells taken from points F, G and H.

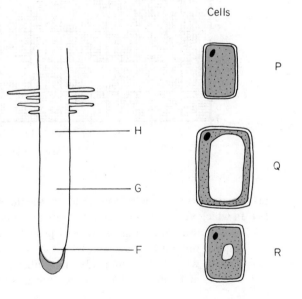

(i) Complete the table to indicate correctly the region of the root tip from which each cell has been taken.

Region of root tip	Letter of cell
F	
G	
H	

(2 marks)

(ii) Further behind the tip, cells become specialised for their functions. What term is used to describe this process? **(1 mark)**

(iii) Name two features, shown in the diagrams of the cells, which are also present in animal cells. **(2 marks)**

(SEB)

Question 8.3

Average mass of humans is shown on the accompanying graph.

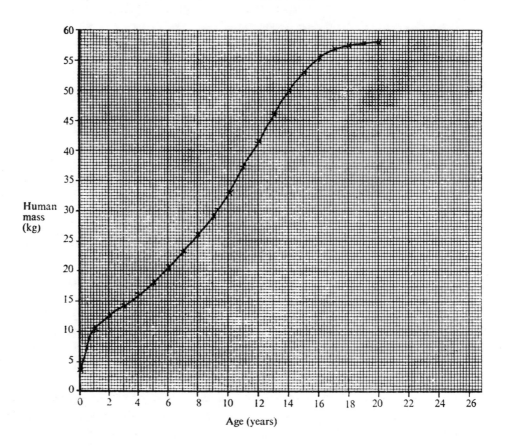

(a) At approximately what age is the average mass 30 kg?
(b) In which year is the growth rate at its maximum?
(c) On the assumption that growth stops at the average age of 20, continue the line on the graph in ink until the age of 24.
(d) Explain why the recommended minimum protein food for humans is greater between the ages of 12 and 17 than between the ages of 17 and 22. **(5 marks)**

(AEB, 1981)

(c) Free-response-type Questions

(i) *Short-answer Questions*

1. State *one* way in which the growth of a plant differs from that of an animal. What part is played in the growth of an animal by: (i) protein; (ii) a named mineral element; (iii) a named vitamin **(4 marks)**

(SUJB)

2. (i) During which stage in the life cycle of a butterfly does the most rapid growth occur?

   (ii) Insects possess an exoskeleton (cuticle). What process occurs at intervals throughout the growth of an animal with an exoskeleton?

   (iii) Which group of food substances is needed by animals for body growth and repair?

   (iv) What term describes the changes from larva to adult stage in a frog?

   (4 marks)

   (SUJB)

3. List, in their correct sequence, five important stages in a twelve-month period in the adult life of a *named* species of deciduous tree. (OLE)

   (ii) *Long-answer Questions*

✳ **Question 8.4**

   (a) Where in a flowering plant would you expect to find the regions associated with its growth in length and its growth in girth? **(4 marks)**

   (b) With the help of diagrams describe how the process of secondary thickening occurs. **(7 marks)**

   (c) Outline the significance of (i) bark, (ii) lenticels, in the life of a tree. **(4 marks)**

   (SUJB)

Answer to Q8.4

This is a mixed question requiring knowledge of plant growth and structure (see Chapter 3).

(a) Regions concerned with growth in *length* are associated with the *apical meristems*, in root and stem tips, and the *elongation* and vacuolation regions, all responsible for *primary* growth.

   Regions concerned with growth in *girth* are associated with the *lateral meristems*, which include the cambium of vascular bundles of stems and roots and are responsible for *secondary* growth.

(b) Secondary *thickening* is another term for secondary growth, which occurs in *woody* flowering plants. The cambium of vascular bundles forms a complete *cambium ring* which then proceeds to form new secondary xylem and phloem. Each year new xylem forms an annual ring. The original central *pith* disappears. Figures 8.7 and 8.15 are needed to illustrate the answer.

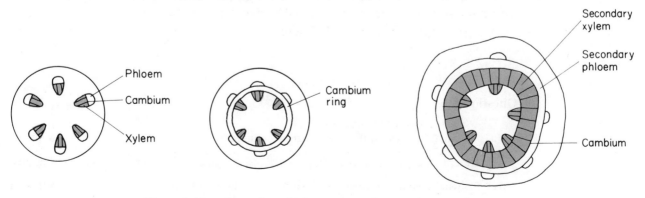

Figure 8.15   Secondary thickening growth stages in a woody stem.

(c) (i) *Bark*, the outermost tissue in woody stems and roots, forms from a *cork* cambium. It is protective, often distasteful to grazing animals, and water-proof to prevent water evaporation.

(ii) *Lenticels* are groups of loosely packed cells seen as pores in woody stems and roots, and allow exchange of gases.

✳ Question 8.5

Give an illustrated description of the structure and germination of a named flowering plant seed. **(25 marks)**

(L)

Answer to Q8.5

This is a common *direct* question. The key words are *illustrated*: diagrams are needed to describe seed *structure* and seed *germination*. The important part of the question asks you to name the flowering plant seed, so do not leave the name out for the examiner to guess.

Select and name the seed, e.g. broadbean, pea, maize, or whichever one you know the structure of.

Seed structure. A sectional view is needed, as in Fig. 8.4(a). List each structure and give its function.

Structure	Function
Testa	Protection (note maize has a 'husk' of combined pericarp and testa)
Micropyle	Water entry
Endosperm	Food reserve — whether endospermic or non-endospermic
Embryo:	
Cotyledons	Either food reserve or future photosynthetic first leaves
Radicle	Future root
Plumule	Future shoot

Seed germination. Here a series of diagrams showing germination of the seed you have selected is required. Explain whether it is showing *epigeal* or *hypogeal* germination. Figure 8.5 is needed here.

✳ Question 8.6

(a) By what features would you recognise an insect? **(6 marks)**

(b) Describe the life cycle of a named *social* insect, e.g. bee, wasp, ant or termite. **(14 marks)**

(c) To what extent can insects be regarded as economically important? **(5 marks)**

(L)

Answer to Q8.6

This is a *mixed* question needing a knowledge of the classification of organisms, Chapter 2, and insect metamorphosis; also knowledge concerning organisms and the environment, Chapter 14.

(a) *Features* that identify insects: body divided into head, thorax and abdomen; three pairs of walking legs; single pair of antennae.

(b) *Life cycle* of the selected insect must include a description of the type of metamorphosis, whether *complete* (with larval forms) or *incomplete* (without larvae).

　　Diagrams must show each stage in metamorphosis. Skill is needed in making a good, *recognisable* diagram.

(c) *Economic importance* means how insects affect human beings and their industrial and commercial activities.
　　(i) *Pests*: insects devour crops and food in store, e.g. locust and flour beetle.
　　(ii) *Transmit disease*: malaria and yellow fever by mosquito; dysentery by housefly.
　　(iii) *Pollinators*: for insect-pollinated plants, e.g. apples and other fruit.
　　(iv) *Biological control* of insect pests can be achieved by using predator insects.

8.4 Self-test Answers to Objective and Structured Questions

Answers to Multiple-choice Objective Questions

1. b　2. d　3. b　4. a　5. a　6. b　7. a　8. b　9. b　10. d　11. a　12. c
13. d　14. c　15. c　16. c　17. a　18. b　19. c　20. d

Answer to Structured Question 8.1

(i) The difference between the dry weight X, 6.5 g − dry weight Y, 5.9 g = 0.6 g, or a *loss in weight* in Y. This is due to the seed food reserve being used to provide energy for growth. This *loss* in weight is shown in the first part of the graph curve in Fig. 8.11.

(ii) The difference in dry weight of Y and Z is:

7.1 g − 5.9 g = 1.2 g

The seeds in group Z have a *greater* dry weight than group Y because they have formed organic substances by photosynthesis, as shown in the second part of the graph curve in Fig. 8.11.

(iii) Two other factors for germination are an optimum temperature (warmth), and air (oxygen).

(iv) *Epigeal* germination: the cotyledons form the first photosynthetic leaves; hypocotyl lengthens. *Hypogeal* germination: the cotyledons remain below ground for food storage and do not act as photosynthetic organs; epicotyl lengthens.

Answer to Structured Question 8.2

(i) F: P, root apex meristem cell
  G: R, vacuole beginning to form
  H: Q, vacuole fully formed in elongation region

(ii) The process is called *cell differentiation*, to form root tissues.

(iii) Two animal cell features shared with plant cells are: the possession of a nucleus, and of cytoplasm (animal cells have very *small* vacuoles).

Answer to Structured Question 8.3

(a) 30 kg mass found at age 9 years.

(b) Growth rate at its maximum from birth to age 1 year; the birth weight *doubles* from about 5 kg to 10 kg in one year.

(c) The line would continue *parallel* to horizontal age axis.

(d) Protein is needed for the considerable growth spurt between ages 12 and 17 years; there is *less* growth between ages 17 and 22 years, therefore less protein is needed for structural growth.

9 Homeostasis

9.1 Theoretical Work Summary

(a) Homeostasis

Homeostasis can be defined as the maintenance of a *constant* internal environment around the cells of mammal tissues, despite changes in *external* environment.

Tissue fluid forms from blood (see section 6.1(g) and Fig. 6.4); its composition or *concentration* of glucose, mineral ions, carbon dioxide, oxygen, water, nitrogenous waste products and pH, together with temperature, must be kept *constant* within normal values.

(b) Homeostatic Control Systems

1. These consist of: sensory *receptors* or *detectors*, detecting changes in tissue fluid components, and may be located in brain, arteries or pancreas.
2. *Feedback* information is sent to the homeostatic control centre, the hypothalamus region of the brain.
3. Control centre activates *effectors*: secretory products from glands which restore the tissue fluid component to its normal value. Main effectors are the pituitary, adrenal glands and pancreas tissue, which produce different hormones causing different chemical changes.

(c) Water and Salts Homeostasis

1. The homeostatic control of water and salts in tissue and blood fluid is called *osmoregulation* or tissue fluid *osmotic pressure* control.

   High osmotic pressure = high salt concentration + low amount of water
   Low osmotic pressure = low salt concentration + high amount of water

2. *Kidneys* are the main homeostatic organs for osmoregulation, maintaining human body water content at 70%, thus preventing *dehydration* due to water shortage (see Figs 9.1 and 9.2).

   Nephrons are the structural and functional units of the mammalian kidney, producing glomerular filtrate by *ultra-filtration* under pressure (Fig. 9.3). *Small* mineral ions or molecules of water, urea (carbamide), glucose and *certain* proteins filter through the glomerulus wall pores. *Reabsorption* of water (osmosis), glucose, *small*-molecule proteins, amino acids, and ions (diffusion and active transport) occurs in convoluted kidney *tubules*, with pH adjustment of tissue fluid.

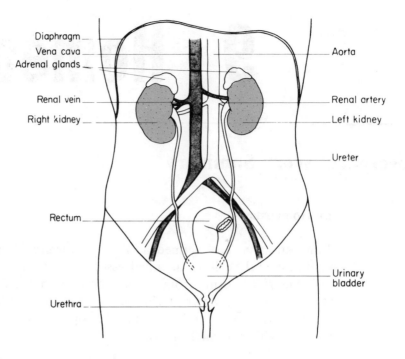

Figure 9.1 The urinary system of a human female. In examinations you will probably be asked to 'label and explain this figure'.

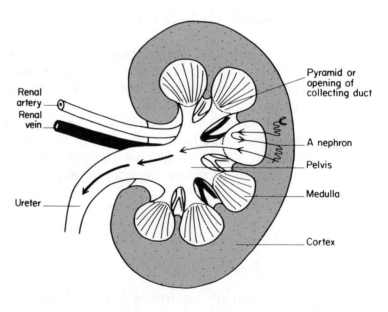

Figure 9.2 Vertical section through a human kidney to show the internal structure. In examinations you will probably be asked to 'label and explain this figure'.

3. *Contractile* excretory *vacuoles* are the organelles for water concentration control (*osmocontrol*) in unicellular plants and animals. (See Figs 2.1(c) and 2.2(a)).

4 The *hormone* ADH, antidiuretic hormone (vasopressin), secreted by posterior *pituitary* gland, causes kidney tubules to *reabsorb* water and decrease volume of urine. This occurs in conditions causing dehydration, e.g. insufficient water intake. Reverse occurs in excessive water intake with *stoppage* of ADH secretion. Homeostatic control is brought about through osmotic pressure detectors with feedback to the osmoregulatory centre in the brain, and effector system via ADH hormone activity.

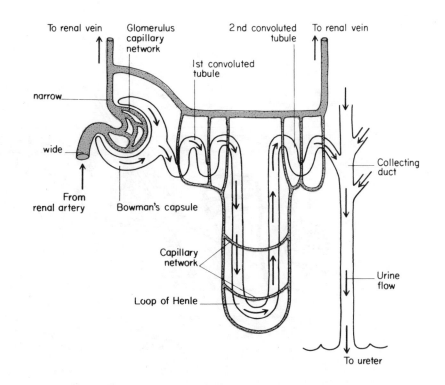

Figure 9.3 The structure of a nephron in the human kidney. In examinations you will probably be asked to 'label and explain this figure'.

5. *Water balance*

Water gains   = Water loss
(by food, drink and metabolic   (by skin evaporation, sweat,
water from respiration)   in faeces and urine)
    Kidney disorders are indicated by presence of *protein* in the urine.

(d) Mammal Body Temperature Homeostasis

1. *Skin* is one of the largest homeostatic organs in mammals.

   It has a *large surface* are, which makes it a good *heat radiator* and *absorber*, and a poor *conductor*. *Sweat glands* produce water, requiring *latent heat* which is drawn from the body for *evaporation*, thereby causing a cooling effect. *Hairs* or *feathers* interlocking trap a heat-insulating layer of air.
2. *Skin structure* is shown in Fig. 9.4. Note the adipose or fat tissue for heat *insulation*; the skin capillary arterioles, which can narrow in *vasoconstriction* to retain heat, and dilate in *vasodilation* to lose heat by radiation; and the erector muscles, which control *interlocking* of hairs.
3. *Endothermic (homoiothermic) birds and mammals* have homeostatic temperature control in brain (hypothalamus). If body temperature alters considerably beyond the *norm*, the control centre brings about changes to restore normal temperature (see Table 9.1).
4. Heat balance in mammals is brought about by the mechanisms summarised in Table 9.2. *Behavioural means* include panting, bathing, wallowing or shade-seeking by overheated animals. Temperature control disorders can cause death through *hypothermia* (body overcooling in infants and elderly), or *hyperthermia* (body overheating in *heat stroke*).
5. *Ectothermic (poikilothermic)* animals are unable to maintain steady body temperatures. Their body temperature changes with that of the external environment. Seen in most animals except mammals and birds.

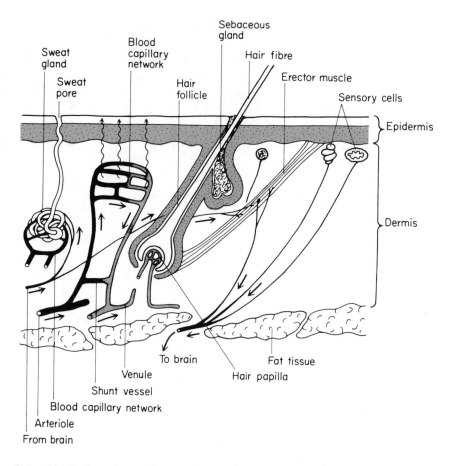

Figure 9.4 Vertical section of human skin to show its internal structure. In examinations you will probably be asked to 'label and explain this figure'.

Table 9.1 Temperature control in birds and mammals

Temperature decreases	Temperature rises
1. Skin arterioles contract	1. Skin arterioles dilate
2. Sweat glands do not function	2. Sweat glands secrete sweat
3. Skeletal muscle action causes shivering, feather and hair erector muscles contract	3. Skeletal muscle relax, feather and hair erector muscles relax
4. Thyroxine increase for sugar respiration and increased metabolic rate	4. Thyroxine decreases leading to decreased metabolic rate

Table 9.2 Heat balance in mammals

Heat gains	Heat losses
1. Aerobic respiration	1. Panting + exhalation
2. Muscle activity	2. Sweat evaporation
3. Liver metabolism	3. Urine, faeces
4. Heat transfer *from* environment	4. Heat transfer *to* environment
5. Hot food ingestion	5. Cold food ingestion

(e) Blood Glucose Homeostasis

1. The *norm* for glucose in blood and tissue fluid is about 0.1%. It is controlled homeostatically by the *liver, pancreas* and *adrenal glands.*
2. The *liver* is a major homeostatic organ in blood composition control. Its cells have a close contact with the blood, as all the blood in the human body passes through the liver every two minutes (see Fig. 9.5). *Excess* glucose is changed into *glycogen* and stored in the liver, and a *deficiency* of glucose in blood causes glycogen to be changed into glucose.

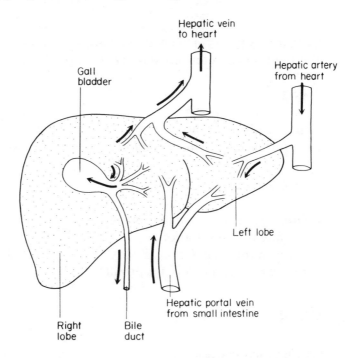

Figure 9.5   Blood supply to the mammalian liver.

3. The *pancreas* is a gland able to detect changes in the concentration of glucose in blood, by means of the *islets of Langerhans* tissue.
    (i) *Increasing* blood sugar concentration causes the pancreas islets to secrete the hormone *insulin*; this causes the liver to change glucose into glycogen. *Sugar diabetes* is evident as the presence of glucose in the urine due to a disorder of the pancreas, which is unable to produce insulin.

$$\text{GLUCOSE} \xrightarrow{\text{INSULIN}} \text{GLYCOGEN}$$

    (ii) *Decreasing* blood sugar concentration causes the pancreas tissue to produce *glucagon* hormone, which, together with the hormone *adrenalin*, changes liver glycogen into glucose.

$$\text{GLYCOGEN} \xrightarrow[\text{ADRENALIN}]{\text{GLUCAGON}} \text{GLUCOSE}$$

    (See also Fig. 10.5.)

(f) Excretion

Excretion is the process of removal of metabolic waste products (water, carbon dioxide, and a mixture of compounds of nitrogen, e.g. ammonia, uric acid and urea (carbamide)). Excretion and osmoregulation are linked to homeostasis. *Aerobic respiration* produces carbon dioxide and water in plants and animals. *Deamination* is the term given to the removal of amino groups from surplus amino acids, producing *ammonia* and/or urea (carbamide) in the vertebrate *liver*. Plants do not perform deamination, and green flowering plants *can use* ammonia, nitrates and urea (carbamide) in protein synthesis.

Excretory Organs

ANIMALS

Excretory organs in animals include the kidneys, lungs and skin sweat glands; also the liver excretes blood breakdown products, i.e. bile pigments.

PLANTS

Plants eliminate carbon dioxide and water via the leaves. Certain flowering plants form nitrogen-containing *alkaloids*, possibly as products of metabolism.

(g) Homeostasis in Flowering Plants

1. *Osmoregulation* of water and salts is under the control of conditions affecting the rate of transpiration.
2. *Temperature regulation.* Plants *absorb* radiant heat energy, and are cooled by *transpiration*; they do not produce much heat by respiration.

9.2 Practical Work

(a) Microscope work

It is important to examine transverse sections of mammal skin, seen either by means of 35 mm transparencies or on prepared microscope slides.

(b) Dissection

A general dissection of the urino-genital system of a frog or small mammal needs to be demonstrated, together with examination of a sheep kidney seen in vertical sectional view.

(c) Cooling Effect of Evaporation

Demonstrated by soaking cotton wool fixed to a laboratory thermometer bulb in water, and in ethanol, and comparing the temperatures recorded when exposed to air, with that recorded by an uncovered thermometer.

(d) Surface Area and Heat Loss

Fill a large flask (1 litre size) and a small flask (25 cm³ size) with boiling water from a kettle. Insert stoppers fitted with thermometers and observe the rate of cooling. The large flask has a *lower* surface area to volume ratio, compared with the smaller flask. This experiment demonstrates the fact that small organisms are liable to lose heat more quickly than large organisms.

9.3 Examination Work

(a) Multiple-choice Objective Questions

1. Urea (carbamide) is made in the mammal:
   (a) kidney
   (b) pancreas
   (c) liver
   (d) stomach

2. The 'master gland' of the mammal endocrine system is the:
   (a) thyroid
   (b) pituitary
   (c) adrenal
   (d) pancreas

3. The gland that produces a hormone which reduces the level of blood glucose is the:
   (a) thyroid
   (b) pituitary
   (c) adrenal
   (d) pancreas

4. Which of the following is the main nitrogenous waste product found in mammalian blood?
   (a) nitrates
   (b) ammonia
   (c) protein
   (d) urea (carbamide)

5. The organism which maintains a steady body temperature of 39°C in spite of changes in temperature of its surroundings ranging from 10 to 35°C would be:
   (a) a fish
   (b) an amphibian
   (c) an insect
   (d) a mammal

6. Which two of the following are a pair of excretory products produced by both plants and animals: (i) oxygen; (ii) water; (iii) mineral salts; (iv) carbon dioxide; (v) urea (carbamide); (vi) ammonia?
   (a) i and iii
   (b) ii and iv
   (c) ii and iii
   (d) i and v

7. Secretion and excretion occur at the same time in:
   (a) exhalation
   (b) sweating (perspiration)
   (c) defaecation
   (d) urination (micturition)

8. Which one of the following is an example of a mammalian excretory product?
   (a) faeces
   (b) urea (carbamide)
   (c) ammonia
   (d) sebum

9. Valuable soluble substances, e.g. glucose, are retained or recovered by the mammalian kidney by the process of:
   (a) ultrafiltration
   (b) active transport
   (c) osmosis
   (d) translocation

10. When a warm-blooded animal (homoiotherm) moves from a warm environment into a colder environment, the following can occur:
   (a) its body temperature rises
   (b) skin blood capillaries become constricted
   (c) skin hair erector muscles relax
   (d) its metabolic rate falls

11. Which *two* of the following are maintained at the correct levels in animals by osmoregulation: (i) protein; (ii) lipids; (iii) water; (iv) blood sugar; (v) salts; (vi) vitamins?
   (a) i and ii
   (b) ii and iv
   (c) iii and v
   (d) iv and vi

12. The main organ of osmoregulation in birds and mammals is the:
   (a) lung
   (b) kidney
   (c) skin
   (d) sweat glands

13. The Bowman's capsule is found in the region of the kidney called the
   (a) pelvis
   (b) ureter
   (c) cortex
   (d) medulla

14. Water constantly enters a single-called freshwater animal such as *Amoeba* by:
   (a) active transport
   (b) filtration
   (c) osmosis
   (d) pinocytosis

15. The structure in mammals which takes urine from the bladder to the exterior is called the:
   (a) urethra
   (b) ureter
   (c) uriniferous tubule
   (d) cloaca

16. The outermost layer of mammalian skin is called the:
   (a) cuticle
   (b) epidermis
   (c) dermis
   (d) ectoplasm

17. The maintenance of a steady temperature and chemical composition of the tissue or extracellular fluid around mammal cells is called:
   (a) excretion
   (b) respiration

(c) homeostasis

(d) homoiothermy

18. Which *one* of the following is *not* under homeostatic control in the human body?
    (a) blood pressure
    (b) blood sugar
    (c) body temperature
    (d) colour vision

19. The main homeostatic control *centre* for mammal body temperature, blood glucose and osmoregulation is located in the:
    (a) brain
    (b) liver
    (c) skin
    (d) kidney

20. The main homeostatic *organ* concerned with osmoregulation in the mammal body is the:
    (a) skin
    (b) liver
    (c) kidney
    (d) pancreas

(b) Structured Questions

Question 9.1

In 24 hours 170 litres of fluid are removed by filtration from the blood by a person's kidneys. In this time only 1.5 litres of urine are formed. Blood plasma contains 0.03% urea (carbamide) in solution; urine contains 0.2% urea (carbamide).
  (i) State *two* functions of the kidney which are indicated by the information above.
  (ii) What is the liquid reabsorbed?
  (iii) What disease is the probable cause of glucose appearing in the urine?
  (iv) What is the cause of this disease?
  (v) Suggest two reasons why mammals possess kidneys and trees do not.   **(6 marks)**
(SUJB)

✳ Question 9.2

When a tadpole undergoes metamorphosis to a frog, there are quite obvious structural changes. The changes of function are less obvious. For example, the tadpole excretes its nitrogenous waste as ammonia whereas the frog excretes urea (carbamide).

(a) Plot a graph of the following data:

Age of tadpole or frog (days)	Ammonia as percentage of total excretory material
50	92
55	88
65	84
75	83
90	68
95	20
100	13
110	12

(7 marks)

(b) The change from excreting mainly ammonia to excreting mainly urea (carbamide) occurs when the frog starts to live on dry land. Between which days do you think the animal leaves the water? **(2 marks)**

(c) Ammonia is very soluble and toxic. Give reasons why it is possible for the tadpole to eliminate it safely whereas it is necessary for the frog to change to eliminating the less toxic compound, urea (carbamide). **(2 marks)**

(d) State *two* of the structural changes which occur in the tadpole to accompany this functional change at metamorphosis. **(2 marks)**

(e) What is the importance of metamorphosis in the frog? **(2 marks)**

(SUJB)

(c) Free-response-type Questions

(i) *Short-answer Questions*

1. Give three ways in which mammals can gain heat. **(3 marks)**

(OLE)

2. What has happened in the body when large amounts of protein appear in mammal urine? **(3 marks)**

(NISEC)

3. State two homeostatic functions of the liver.

(ii) *Long-answer Questions*

* Question 9.3

Explain how mammals and flowering plants eliminate the waste products of their metabolism. **(25 marks)**

(L)

Answer to Q9.3

Products of Metabolism

MAMMALS

1. Respiratory products: carbon dioxide and water.
2. Excretory products: mainly *salts*: chloride, sulphate, phosphate and *nitrogenous* urea (carbamide).

FLOWERING PLANTS

1. Respiratory products: carbon dioxide and water.
2. Nitrogen-containing substances called *alkaloids*, e.g. nicotine; other substances include calcium oxalate crystals which *may* be end-products of metabolism or waste products stored in the plant. Otherwise nitrogen, as nitrates with other salts, is used to form protein and is essential for plant structure and function.

Methods of Elimination

MAMMALS

1. Kidneys: excrete excess water, salts and urea (carbamide).
2. Lungs: water and carbon dioxide.
3. Skin: water, and small amounts of salts and urea (carbamide).
4. Liver: bile pigments from breakdown of haemoglobin.

FLOWERING PLANTS

1. Leaves: transpire water and lose some carbon dioxide and oxygen. *Leaf-fall* may be regarded as a method of elimination of materials no longer needed by the flowering plant.
2. Lenticels.
3. Root hairs eliminate some carbon dioxide.

✳ Question 9.4

About 70% of the body weight of an adult human being is made up of water and the amount remains constant.

(a) List the ways in which water is gained and lost by the mammalian body.   **(3 marks)**

(b) Describe how the concentration of the blood is kept constant within narrow limits.
                                                                                      (8 marks)

(c) What structures in a leaf are associated with control of water loss and how do they function?   **(4 marks)**
                                                                                      (SUJB)

Answer to Q9.4

(a)

Water gains	Water losses
Food	Skin epidermis evaporation
Drink	Skin sweat glands
Metabolic water, product of respiration	Urine

(b) Blood *concentration* means the amount of dissolved substances, e.g. glucose, salts, dissolved gases, urea (carbamide), pH and the amount of water. *Homeostasis* is the method of maintaining a constant composition of blood and tissue fluid, by homeostatic organs. The *kidney* controls water, pH, urea and salt content, and the *glucose content* is controlled by the pancreas and also by the liver and the thyroid and adrenal glands.

(c) Water loss from a leaf is by *transpiration*, a process of evaporation described in Chapter 6. The *stomata* are the structures that control water loss. Stomata generally open in light and close in the dark. Light causes the change of starch made by guard cell chloroplasts into glucose. Glucose formation causes guard cells to become turgid and the stoma pore opens.

✳ Question 9.5

    (a) Draw a large labelled diagram of the mammalian skin to show its structure.   **(8 marks)**

    (b) Define the term 'homoiothermic' and state two advantages of being homoiothermic.
                                                                           (4 marks)

    (c) List four physiological methods which function to prevent overheating of a mammallian body.   **(8 marks)**

Answer to Q9.5

(a) Figure 9.4 is needed here.

(b) *Homoiothermic* (also called endothermic) means the maintenance of a constant body temperature irrespective of environmental temperature changes. Birds and mammals have the advantage of being able to *adapt* to different climates and maintain *active* lives; this assists in their survival.

(c) Physiological methods to prevent the mammal body overheating include:
    (i) sweat evaporation
    (ii) vasodilation of skin arterioles
    (iii) relaxation of hair erector muscle, allowing heat loss by radiation
    (iv) panting to encourage water evaporation from tongue and lung surfaces.

9.4 Self-test Answers to Objective and Structured Questions

Answers to Multiple-choice Objective Questions

1. c   2. b   3. d   4. d   5. d   6. b   7. b   8. b   9. b   10. b   11. c   12. b
13. c   14. c   15. a   16. b   17. c   18. d   19. a   20. c

Answer to Structured Question 9.1

  (i) Two homeostatic functions of the kidney are:
    (a) maintain the correct level of *water* in the blood,
    (b) remove excess urea (carbamide) and maintain low concentrations of this nitrogenous waste in the blood.

  (ii) The main liquid reabsorbed by the kidney is water.

  (iii) Glucose appears in the urine when a person has sugar diabetes.

  (iv) The cause of the disease is a defective pancreas and inability to produce insulin or glucagon.

  (v) Animals have a higher intake of nitrogenous materials in protein foods, surplus amino acids are deaminated and the urea formed is removed by the kidney. Plants convert nitrogen into proteins by protein synthesis; surplus protein is *stored*.

      Trees have a high water intake via a great surface area of absorption provided by root hairs and roots, which is transpired by an equally large surface area of leaves. The mammal kidney can cope with the modest daily intake of water in food and drink.

Answer to Structured Question 9.2

(a) The graph is shown in the accompanying diagram.

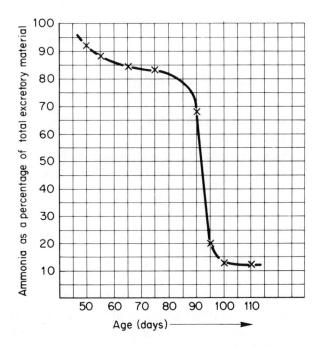

(b) The animal must leave the water between days 80 and 95 when the amount of ammonia in the excretory product drops from 80% to 15%.

(c) The highly soluble ammonia is removed rapidly by the large amounts of water entering the tadpole body across its skin, when it is submerged in water, to leave in large volumes of diluted urine. Frogs on land no longer have the continuous intake of large amounts of water, and therefore the ammonia must be changed into less toxic urea and voided in more concentrated, limited amounts of urine.

(d) Two structural changes that accompany the excretory functional changes are the development of the kidneys and of the urinary bladder.

(e) Metamorphosis is an important *adaptation* for a change from life in water to one on land.

10 Irritability

10.1 Theoretical Work Summary

(a) Constant Environment

Living organisms prefer a constant environment *internally*, around their cells, or *externally*, around their bodies. The internal *control* of this environment is called *homeostasis*.

(b) Stimuli

A stimulus is a change in a component of either the internal or external environment.
1. *External environment components* include heat, light, sound, pressure, gravity, chemicals, water, food, and other living organisms.
2. *Internal environment components* include water, glucose, mineral ions, pH and temperature or *tissue fluid* components.

Receptors

These are the cells or organs that receive the stimuli.
1. Flowering plants have receptors in stem and root *apices*.
2. Mammals have *internal* receptors in muscles and arteries, and *external* receptors in skin, eyes, nose, tongue and ears (sense organs).
   (a) *Light* receptors in eye.
   (b) *Chemical* receptors in nose and tongue.
   (c) *Sound* receptors and balance detectors in ear.
   (d) *Temperature*, *pain* and *touch* pressure receptors in skin (see Fig. 9.4).

(c) Response

A response is any change in an organism made in reaction to the stimulus.
1. *External* environment stimuli cause the organism to move away or towards the stimulus (plant *tropisms*, and animal *movement* and *locomotion*).
2. *Internal* environment stimuli cause the organism to alter the stimulus (e.g. mammal tissue fluid *homeostasis* (see Chapter 9).
3. Animal responses are *rapid*. Plant responses are *slow*.

Effectors

Specialised cells and organs act as effectors in animals and include *muscles*, *cilia*, *flagella*, and *exocrine glands* secreting by ducts on to a surface, reacting to *external* environment stimuli; ductless *endocrine* glands produce hormones directly into the bloodstream, reacting to *internal* environment or tissue fluid stimuli.

Flowering plants have effectors in the form of *elongation-region* cells, which vacuolate under the influence of plant hormones.

(d) Coordination in Mammals

Receptors and effectors are linked by the *nervous* and *hormonal* system, or one *co-ordinating* system, the *neuro-endocrine* system.

1. The *nervous system* consists of *neurones* or nerve cells (Fig. 10.1). The *sensory neurones* connect the external or internal receptors to the *brain* control centre,

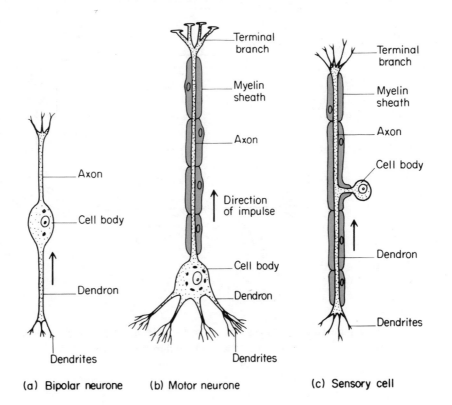

(a) Bipolar neurone   (b) Motor neurone   (c) Sensory cell

Figure 10.1   Three different types of neurones (nerve cells) found in mammals.

or with *relay* neurones in the spinal cord, and *motor neurones* connect with the effectors. The pathway is called a *reflex arc* (Fig. 10.2).

2. The *hormonal system* consists of different ductless *endocrine glands*, directly under the control of the *pituitary* 'master gland', linked to the brain *hypothalamus* control centre.

   Internal receptors feed back information concerning changes in composition of tissue fluid to control centres which cause different effectors to alter tissue fluid composition and restore internal homeostasis (see Chapter 9). Endocrine glands adjust their secretions mainly by the effect of the *amount* of their own hormones circulating in the blood. This is done by *feedback* (see Figs 10.4 and 10.5).

(e) Mammalian Central Nervous System

The mammalian central nervous system consists of:

1. *Brain* (Fig. 10.6) with cranial nerves:
   (a) *Fore-brain, cerebrum* — the coordination centre, with sensory and motor areas; *hypothalamus*, for internal environment control; *olfactory lobes*, which detect smell.

173

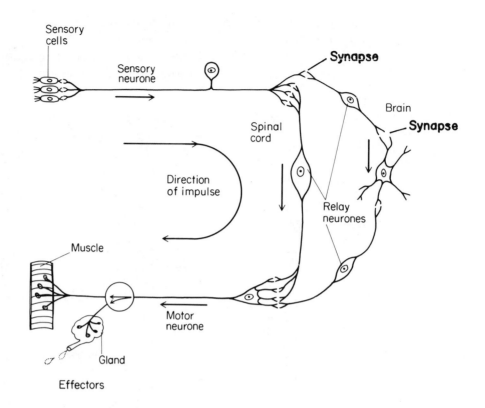

Figure 10.2 Arrangement of nerve cells in a reflex arc. In examinations you will probably be asked to 'label and explain this figure'.

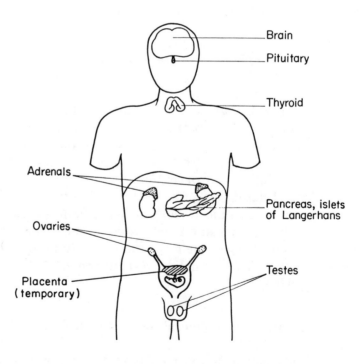

Figure 10.3 Position of the main endocrine or ductless glands in the human body. In examinations you will probably be asked to 'label and explain this figure'.

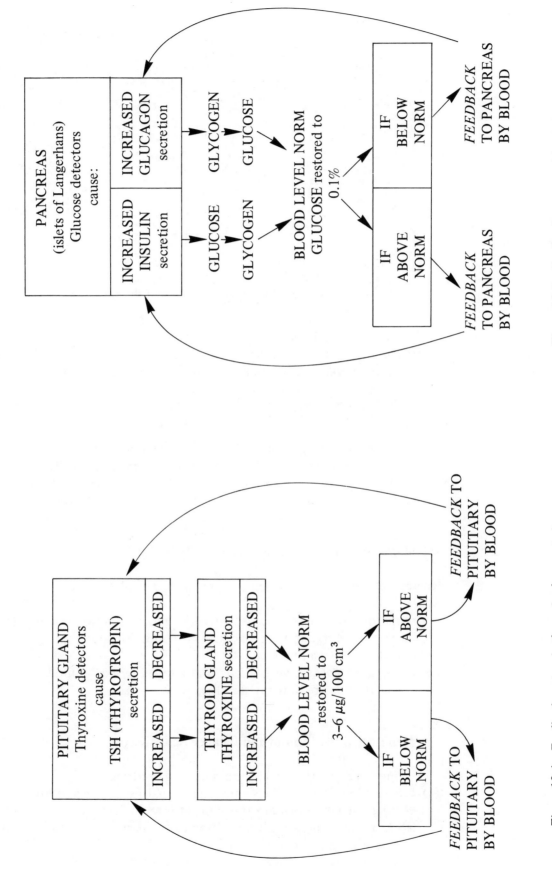

Figure 10.5 Feedback system in blood glucose homeostasis.

Figure 10.4 Feedback system in thyroxine homeostasis.

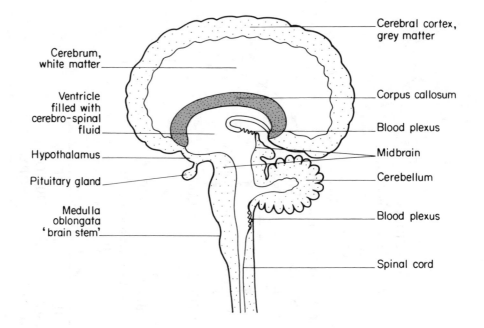

Figure 10.6 Vertical section of the human brain.

(b) *Mid-brain* with *optic* lobes, for sight.
(c) *Hind-brain* (*cerebellum*), for balance and muscular coordination; and the *medulla oblongata*, which controls breathing and heart rate.
2. *Spinal cord* with spinal nerves.
   (a) Dorsal root carries fibres of *sensory* nerves.
   (b) Ventral root carries fibres of *motor* nerves.

   The *grey matter*, composed of nerve cell bodies, constitutes the *outer* parts of the cerebrum and the *inner* parts of the spinal cord. *White matter* is composed of nerve axon fibres, and forms the *inner* parts of the cerebrum and the *outer* parts of the spinal cord.

(f) Mammalian Eye: Light Sensitivity

The eye of a mammal is a sense organ with the following characteristics (see Fig. 10.7):
(a) *retina* of light-sensitive cells, the *rods*, which function in dim light and detect black and white; and the *cones*, which function at higher light intensities and detect red, blue and green;
(b) transparent *cornea*, with a *lens* that adjusts by accommodation for *focusing*. *Eye defects* are illustrated in Fig. 10.8.

(g) Mammalian Ear: Sound and Balance Sensitivity

The mammalian ear (Fig. 10.9) is a sense organ for:
(a) *balance*: by means of the semicircular canals, sacculus and utriculus;
(b) *hearing*; air vibrations are transmitted into the cochlea, where cells are stimulated, and the impulse is transmitted to the brain by the auditory nerve.
Hearing defects can be due to:
(a) blockage — wax in the external auditory meatus
(b) burst or perforated eardrum

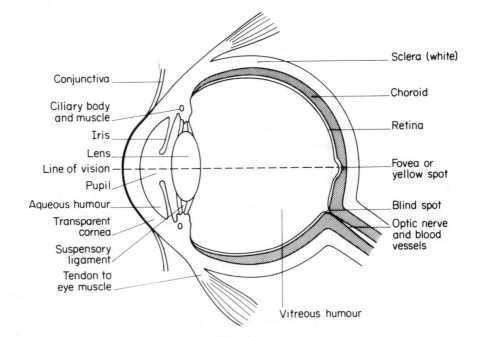

Figure 10.7 Vertical section of the human eye to show its internal structure. In examinations you will probably be asked to 'label and explain this figure'.

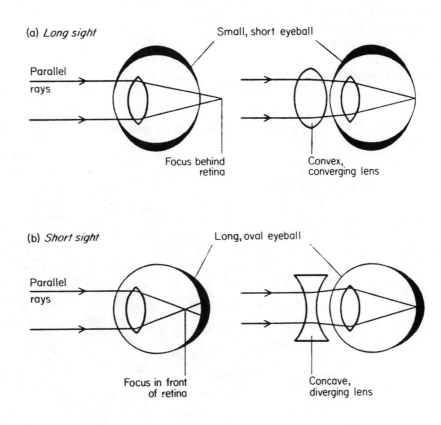

Figure 10.8 Eye defects and their correction by lenses.

(c) bone growth and fusion of the ossicles (the malleus, incus and stapes), and closure of the round and oval windows

(d) nerve destruction

(h) Mammalian Hormone Systems

Mammalian hormones and effects are summarised in Table 10.1.

Table 10.1   Effects of mammalian hormones

Gland and hormone	Excess effect	Deficiency effect
Pituitary near to hypothalamus of brain:		
growth hormone	Gigantism	Dwarfism
trophic hormone	Stimulates other endocrine glands	Retards other endocrine glands
Antidiuretic hormone (ADH, vasopressin)	Urine concentrated and scanty	Urine dilute and copious
Thyroid: thyroxine	Increased basal metabolic rate, decreased body weight, mental excitability, fast pulse	Decreased basal metabolic rate, increased body weight, mental sluggishness, slow pulse
Pancreas: insulin and glucagon	Nervous convulsions, low blood sugar level	Sugar glucose in urine — diabetes, high blood sugar level
Testes: male sex hormones. Secondary sexual characteristics, i.e. facial and pubic hair, voice change		
Ovaries: female sex hormones. Secondary sexual characteristics, i.e. mammary gland growth, pubic hair, menstrual-cycle regulation		

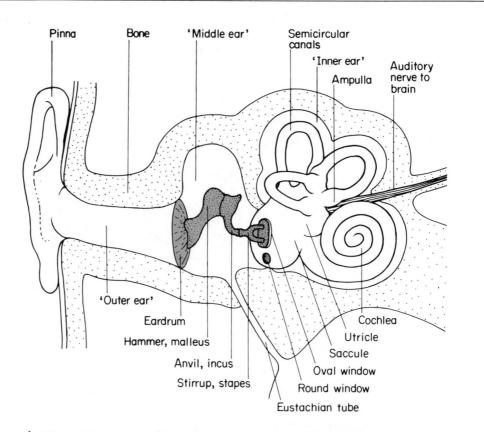

∗ **Figure 10.9**   Simplified diagram to show the structure of the human ear.

(i) Animal Behaviour

Behaviour is the *observable activity* of a *whole* animal in response to stimuli. There are two categories, innate and learned (see Table 10.2).

Table 10.2   A comparison of innate and learned behaviours

Innate behaviour	*Learned behaviour*
Inborn responses to external stimuli. Same in all members of a species	New or modified patterns of behaviour gained by experience. Different in members of a species
Simple reflex action, e.g. touch of food on tongue causing salivation	Conditioned reflex action, e.g. sound of bell at feeding time causing salivation

Taxis is the behaviour of simple unicellular plants or animals in the form of movement of the *whole* body in response to different stimuli.

10.2 Practical Work

(a) Eye Blind Spot

This is the retina region where the optic nerve enters the eyeball; it is *without* rods and cones.

+   ●

With the left eye closed, concentrate the vision of your right eye on the cross. Slowly bring the page towards your face. The dot disappears when its image falls on the retina blind spot.

(b) Perception of Stimuli by Skin Receptors

(i)   Blindfold a person.

(ii) *Heat*-sensitive receptors of the forearm skin are located using a knitting needle heated in hot water. Mark skin with red ink.

(iii) *Cold* receptors of the skin are located using a knitting needle cooled in a freezing mixture of crushed ice and salt. Mark skin with blue ink.

(iv) *Pressure or touch* spots are located with a stiff hogshead bristle or nylon filament, mounted on a card strip or held in forceps.

10.3 Examination Work

(a) Multiple-choice Objective Questions

1. The photoreceptors of the eye are found in the:
   (a) sclera
   (b) choroid

(c) cornea

(d) retina

2. The intensity and amount of light reaching the inner coat of the eye is controlled by the:
   (a) conjunctiva
   (b) iris
   (c) cornea
   (d) lens

3. The shape of the eye lens is controlled by the:
   (a) aqueous humour pressure
   (b) vitreous humour pressure
   (c) iris sphincter muscles
   (d) ciliary muscle

4. Most of the refraction of light occurs at the:
   (a) cornea
   (b) lens
   (c) aqueous humour
   (d) vitreous humour

5. Vitamin A (retinol) is necessary for the pigment found in the retina
   (a) red light cones
   (b) blue light cones
   (c) green light cones
   (d) rods

6. The eustachian tube connects the pharynx of mammals with the:
   (a) auditory nerve
   (b) middle ear
   (c) inner ear
   (d) outer ear

7. Sound receptors are situated in the ear
   (a) utricle (utriculus)
   (b) saccule (sacculus)
   (c) cochlea
   (d) semicircular canals

8. When sound waves from the air enter the ear, the first membrane that they strike and vibrate is called the:
   (a) round window
   (b) tympanic membrane
   (c) oval window
   (d) membranous labyrinth

9. The semicircular canals of the mammal ear make the animal aware of:
   (a) sound pitch
   (b) sound loudness
   (c) head movements
   (d) body posture

10. The ear ossicles are three small bones of the mammal ear connecting the
    (a) oval window with the eardrum
    (b) round window with the oval window
    (c) round window with the eardrum
    (d) ear pinna with the eardrum

11. The endocrine gland found attached close to the brain hypothalamus is the:
    (a) thyroid
    (b) adrenal
    (c) pituitary
    (d) thymus

12. The receptors found in the eye are called:
    (a) interoreceptors
    (b) exteroreceptors
    (c) proprioreceptors
    (d) thermoreceptors

13. The white matter of the brain and spinal cord is composed mainly of nerve cell
    (a) axon fibres
    (b) cell bodies
    (c) dendrons
    (d) terminal branches

14. Which of the following is called a conditioned or learned reflex?
    (a) salivation at the sound of a dinner gong
    (b) salivation when food touches inside the mouth
    (c) withdrawal of a hand from a hot surface
    (d) knee-jerk on striking a tendon below the knee-cap

15. A synapse is a point of contact of the terminal branch of a neurone made between which *two* of the following: (i) muscle membrane, (ii) neurone cell body; (iii) neurone dendrite; (iv) neurone axon; (v) neuron myelin sheath; (vi) adipose fat cell?
    (a) i and iii
    (b) ii and iv
    (c) ii and v
    (d) iii and vi

16. The region of the brain which shows greater development in human beings compared with other mammals is the:
    (a) cerebrum
    (b) cerebellum
    (c) medulla oblongata
    (d) thalamus

17. Hormones pass from the endocrine glands to regions where they effect changes by or through:
    (a) tubular ducts
    (b) blood vessels
    (c) nerve fibres
    (d) vascular bundles

18. The part of the brain concerned with maintaining balance is the:
    (a) cerebrum
    (b) cerebellum
    (c) medulla oblongata
    (d) thalamus

19. The chemical elements present in thyroxine are:
    (a) thorium, hydrogen, oxygen, nitrogen and iron
    (b) carbon, hydrogen, oxygen, sulphur and iodine

(c) calcium, hydrogen, oxygen, nitrogen and indium

(d) carbon, hydrogen, oxygen, nitrogen and iodine

20. Which one of the following gives the correct flow of a nerve impulse in a mammalian nerve cell reading from left to right?
    (a) cell body – dendron – terminal branch – axon
    (b) dendron – cell body – axon – terminal branch
    (c) axon – dendron – cell body – terminal branch
    (d) terminal branch – axon – cell body – dendron

(b) Structured Questions

✳ **Question 10.1**

Time (days)	Pulse rate per minute	Mass (kg)	Basal metabolic rate (% normal)
0	70	66	55
2	70	66	55
4	–	62	82
6	88	63	85
8	82	–	–
10	80	60.5	105
14	–	60	–
16	80	–	98
22	66	59	80

The table shows the effects of administration of three doses of thyroxine (thyroid hormone) to a person suffering from a thyroid deficiency. Where no figures are shown on any day, this signifies no record was taken.

(a) Plot suitable graphs of the data.   **(8 marks)**

(b) On what day was the patient's basal metabolic rate at a normal value?   **(1 mark)**

(c) What element is needed to produce thyroid hormone?   **(1 mark)**

(d) Deduce from the data and graphs some of the symptoms you would expect to encounter in a person suffering from an excess secretion of thyroid hormone.   **(5 marks)**

(SUJB)

✳ **Question 10.2**

Irritability or 'response to a stimulus' is shown in the following examples:
Plant example: the roots of a seedling placed on its side grow and bend through 90° and then continue growing downwards.
Mammal example: in bright light the pupil of the eye becomes constricted.

(a) Divide the page vertically into two halves, one headed *plant example*, and the other headed *mammal example*. Answer the following questions for each example.
   (i) What is the stimulus?   **(2 marks)**

   (ii) What is the region or tissue that forms the receptor for the stimulus?   **(2 marks)**

   (iii) What is the response?   **(2 marks)**

(iv) What is the effector region or tissue carrying out the response? **(2 marks)**

(v) How does the effector make the response? **(6 marks)**

(vi) What is the approximate time interval between first receiving the stimulus and making the response? **(2 marks)**

(vii) Is the response permanent or temporary? **(2 marks)**

(viii) What changes within the organism link the stimulus and response? **(8 marks)**

(b) 1. Select two of your answers to (a) which are true for all stem and root tropisms. **(2 marks)**

2. Select two of your answers to (a) which are true for all mammalian reflex actions. **(2 marks)**

(AEB, 1981)

Question 10.3

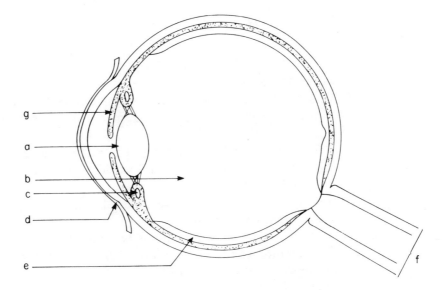

The accompanying diagram represents a vertical section through a mammalian eye. Name parts a, b, c and d. State ONE function of each of the following: e, f and g. **(5 marks)**

(NISEC)

(c) Free-response-type Questions

Short-answer Questions

1. What optical defect is corrected by spectacles with concave lenses? **(1 mark)**

What would have been the symptoms of the optical defect if convex lenses had been prescribed? **(1 mark)**

(OLE)

2. List the important parts of the nervous system through which nerve impulses must pass between a stimulated sense cell of the hand and the muscle that can cause the hand to be jerked away. **(5 marks)**

(OLE)

3. Give the names of the parts of the ear which have the following functions:
   (a) Equalise pressure on either side of ear drum.
   (b) Detect turning movements of head.
   (c) Passes vibrations across middle ear.
   (d) Two main parts of ear normally filled with air.   **(4 marks)**
                                                            (SEB)

Long-answer Questions

* **Question 10.4**

   (a) Construct a large labelled diagram of the eye of a mammal seen in section.   **(8 marks)**

   (b) Name three structures through which light will pass and state the function of each.
                                                                               (4 marks)

   (c) Briefly contrast rods and cones.   **(3 marks)**
                                                            (SUJB)

Answer to Q10.4

(a) This is a common question, requiring ability to make a neat labelled diagram of the eye from memory. A diagram similar to Fig. 10.7 is drawn here.

(b) Light passes through:
   (i) *cornea*, which causes greatest refraction or bending of light;
   (ii) *aqueous humour*, which provides nutrients to the cornea cells, and moisture for gas exchange;
   (iii) *lens* has variable powers of refraction controlled by ciliary body and suspensory ligaments.

(c) *Rods* are sensitive to low-intensity light; there are many in nocturnal animals who require good night vision. *Cones* are sensitive to bright light, and are responsible for colour vision.

* **Question 10.5**

   (a) Make a large labelled diagram of the mammalian ear to show its internal structure.
                                                                               (8 marks)

   (b) Choose five of the structures you have labelled and state their functions.   **(5 marks)**

   (c) How would a mammal that became totally deaf in one ear be at a disadvantage compared with those with normal hearing?   **(2 marks)**
                                                                               (SUJB)

Answer to Q10.5

(a) A neat labelled diagram of Fig. 10.9 is needed here.

(b) Any *five* labelled parts are selected and their functions are stated.
   (i) *Pinna* is funnel shaped to direct sound waves into and along the auditory canal.
   (ii) *Eardrum* vibrates with sound waves.
   (iii) *Ossicles*, three small bones, transmit and amplify sound vibrations.

(iv) *Cochlea* has sound receptors.

(v) Semicircular canals have receptors detecting turning movements of head.

(c) A mammal deaf in one ear would be unable to locate and detect sounds from side of body with deaf ear. It would also need to turn head frequently to locate and hear sounds.

✳ Question 10.6

Give an account of the effects of hormones in flowering plants and mammals.　　**(25 marks)**

(L)

Answer to Q10.6

This question has the key word *effects*, and consequently a description of glands and sources is not required. The answer can be arranged in tabular form.

Flowering plant hormones	*Mammalian hormones*
1. The main effect is on plant *growth*	1. The main effect is on the control of *internal homeostasis* or tissue fluid composition
2. The plant hormones are produced in small amounts in stem and root apices	2. Lesser effect is in growth control by the growth hormone of pituitary gland, and by the action of the thyroid gland
3. The main effect is cell *elongation* as brought about by auxins and gibberellins; cell *division* is promoted by cytokinins	3. Changes in the internal environment or tissue fluid composition are regulated mainly by 'feedback' of information by hormones. ADH controls water levels. Glucagon and insulin control glucose levels, and thyroxine the basal metabolic rate
4. High concentration of plant hormones in weedkillers produce abnormal growth and plant death	4. *Reproduction* is controlled by pituitary gland hormones and the hormones from the testes and ovaries
5. *Growth movements* are controlled internally by auxins, producing various phototropic and geotropic responses to external stimuli	5. Animal hormones, transported rapidly by the blood, are an important means of body cell *communication*

10.4　Self-test Answers to Objective and Structured Questions

Answers to Multiple-choice Objective Questions

1. d　2. b　3. d　4. a　5. d　6. b　7. c　8. b　9. c　10. a　11. c　12. b
13. a　14. a　15. a　16. a　17. b　18. b　19. d　20. b

Answer to Structured Question 10.1

(a) The graphs for the data are shown in the diagram.

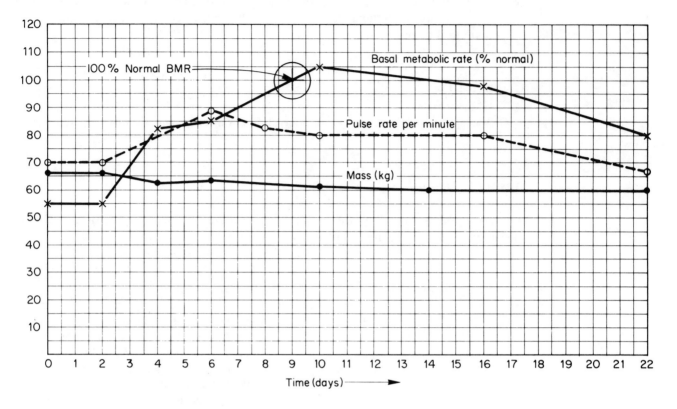

(b) Basal metabolic rate normal on day 9.

(c) Iodine.

(d) Symptoms of *excess* thyroid secretion include: body weight *loss*; high pulse *rate*; higher metabolic rate.

Answer to Structured Question 10.2

This is a 'mixed' question requiring knowledge of plant growth movements (Chapter 8) and the mammalian nervous system. The answer is set out in columns, as asked in the question: see next page.

Answer to Structured Question 10.3

Names of labelled parts: a, lens; b, vitreous humour, c, ciliary body or muscle, d, conjunctiva.

Function of labelled parts: e, *retina* (layer of rods and cones, which are photo-receptors, i.e. light receptors); f, *optic nerve* leads to the brain, making animal aware of its external environment, and assists in balance and posture; g, *iris* regulates amount of light entering the lens and reaching the retina.

Question	Plant example	Mammal example
(a)		
(i) What is the stimulus?	Gravity	Light intensity
(ii) Receptor	Root apical meristem	Retina sensory cells
(iii) Response	Curvature	Pupil contracts
(iv) Effector	Elongation-region cells	Iris sphincter muscles
(v) Effector mechanism	Cells vacuolate causing unequal growth	Nerve impulse causes iris muscles to contract
(vi) Time	Long duration – hours	Rapid – seconds
(vii) Permanent or temporary	Temporary	Temporary
(viii) Changes linking stimulus and response	Auxin hormone	Nervous impulse
(b)	1. True for all stem and root tropisms: iii, iv, v, vi and viii	2. True for all mammalian reflex actions: v, vi, vii and viii.

11 Support and Movement

11.1 Theoretical Work Summary

(a) Support in Plants

1. *Unicellular* plants (algae and fungi) have:
   - (i) firm *cell walls* of cellulose (algae) or fungal cellulose and chitin in fungi;
   - (ii) *cell turgor*, due to osmotic pressure on cell walls, which gives rigidity and support (see Fig. 6.3 and section 6.1(f)).
2. *Multicellular* flowering plants have:
   - (i) *cell walls* strengthened by non-living *cellulose* and *lignin*;
   - (ii) *cell turgor*: see Fig. 6.3 and Table 6.4;
   - (iii) *supporting tissue*, i.e. collenchyma, sclerenchyma and vascular bundles in primary unthickened stems (*lignified* tissue is found in sclerenchyma and secondary thickening or *wood* of trees and shrubs: see Fig. 3.2, section 8.1(f), and Fig. 8.7;
   - (iv) *climbing plants*, e.g. honeysuckle and runner bean, have *twining stems*; vine and sweet pea have *tendrils*; and blackberry has *prickles*.

(b) Movement in Plants

1. *Unicellular* plants, e.g. algae such as *Chlamydomonas* and *Euglena*, have flagella which enable *taxic* movement to occur.
2. *Multicellular* flowering plants are usually rooted and static. The movements are mainly *tropisms* of stem, root or leaf, caused by change in *cell turgidity* by *vacuolation*, through effect of plant hormones (Section 8.1(i) 5).

(c) Support in Animals

1. *Unicellular* animals, e.g. *Amoeba*, have a sensitive, contractile plasma membrane which causes a change in body shape in response to *stimuli*.
2. *Multicellular* animals have skeletons:
   - (i) The *hydrostatic skeleton* in earthworm consists of *contractile muscles* acting against *incompressible body fluid* (Fig. 11.1).
   - (ii) An *exoskeleton* is a *protective shell* in insects, crustaceans and molluscs, surrounding the body and periodically removed by moulting (*ecdysis*). *Inflexible* shell or cuticle parts are separated by *flexible* parts or joints. *Muscles* are connected internally by connective tissue to the rigid exoskeleton.
   - (iii) The *endoskeleton* in vertebrates consists of a living rigid skeletal tissue, i.e.

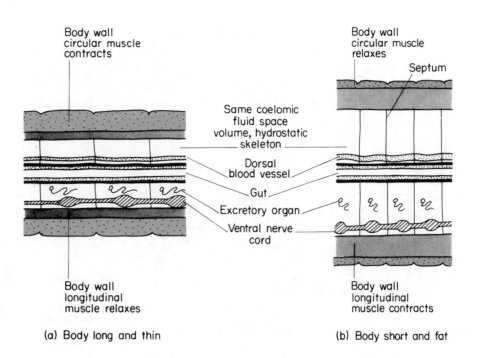

(a) Body long and thin

(b) Body short and fat

Figure 11.1   Antagonistic muscle action in the earthworm, *Lumbricus* sp.

bone and cartilage, and contractile skeletal muscle tissue, connected together by connective tissue, ligaments (which join bone to bone) and tendons (which join muscle to bone).

(d) Mammal Skeleton

1. The vertebrate skeleton is composed of:
    (i) *axial skeleton*, i.e. the skull with cervical, thoracic, lumbar, sacral and caudal *vertebrae*;
    (ii) *appendicular skeleton*, composed of pelvic and pectoral *girdles*, with fore/upper and hind/lower limbs (Fig. 11.2).

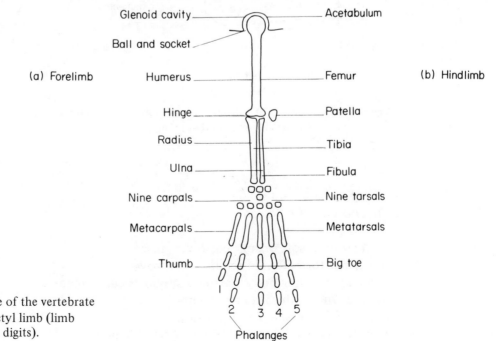

Figure 11.2   Structure of the vertebrate pentadactyl limb (limb with five digits).

2. *Joints* are where two or more bones meet:
  (i) *immovable* or fibrous joints (between skull bones);
  (ii) *slightly movable* joints joined by cartilage (between vertebrae);
  (iii) *freely movable synovial* joints (i.e. hinge and ball-and-socket joints; see Fig. 11.3).

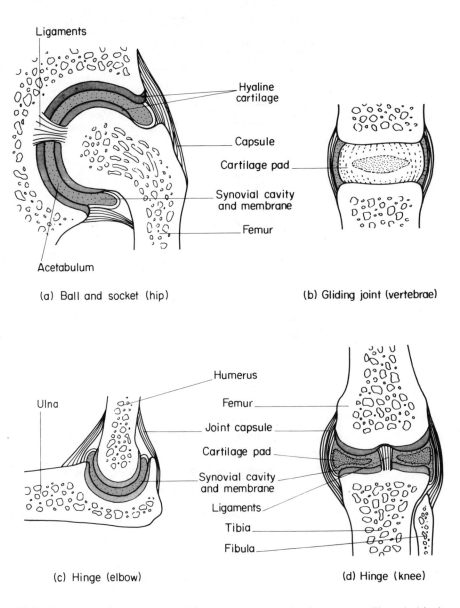

Figure 11.3 Structure of various movable joints. In examinations you will probably be asked to 'label and explain these figures'.

Functions of the Skeleton

Protection of soft organs, e.g. lungs, heart
Blood cell formation in red bone marrow
Support in raising body; cartilage supports ear pinnae
Respiration in rib cage movement
Ear ossicles for *sound* transmission
Locomotion or *movement* of part or whole body

190

(e) Effectors

1. *Muscles* and *exocrine* or ducted *glands* are effectors responding to nervous stimuli.
2. Muscles are *contractile* tissues and are of three types:
    (i) *skeletal*, which is also known as voluntary or striped, is connected to endoskeleton (vertebrates) or exoskeleton (arthropods), and causes body movement and locomotion;
    (ii) *smooth* (involuntary or unstriped), found in the gut wall, causes *peristalsis*;
    (iii) *cardiac* or heart muscle pumps blood fluid.
3. *ATP (adenosine triphosphate)* causes muscle tissue to *contract*.
4. *Antagonistic muscles* are pairs of muscles producing opposite effects, one contracting while the other relaxes.

Sphincter muscles are circular muscles surrounding a tube or opening, e.g. mouth, anus, bladder, eye, iris and stomach pylorus.

(f) Limb Movements

Locomotion is a movement of the *whole* animal body from place to place brought about by limb movements following skeletal muscle contraction.

Limbs are *levers* (Fig. 11.4) composed of rigid bones, turning about a joint or *pivot* by means of effort from muscle contraction, resulting in the movement of the body load.

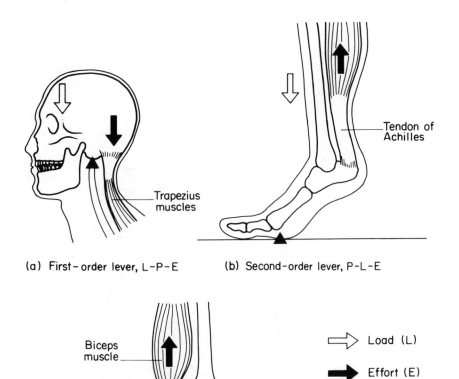

(a) First-order lever, L-P-E

(b) Second-order lever, P-L-E

(c) Third-order lever, L-E-P

Tendon of Achilles

Trapezius muscles

Biceps muscle

Load (L)

Effort (E)

Pivot (P)

Figure 11.4   Limbs as levers.

(i) *Insect Limb Movement and Locomotion*

Antagonistic muscle pairs connected to inside of *exoskeleton* act across *flexible joints* in exoskeleton by alternate contraction of flexor and extensor muscles (see Fig. 11.5).

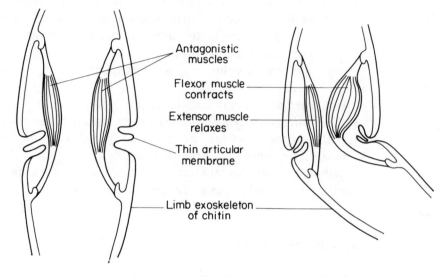

Figure 11.5 Antagonistic muscles in an insect leg.

(ii) *Mammal Forelimb Movement*

1. The *biceps* has *two* points of origin on a fixed bone (the scapula); it is the *flexor* muscle to *bend* the limb. Biceps contraction raises the lower arm through effort acting on an *insertion* or attachment point to the movable bone (the radius).
2. The *triceps* has *three* points of origin on fixed bones (the humerus and scapula); it is the *extensor* muscle to *straighten* the limb. Triceps contraction effort acts on *insertion* to movable bone (the ulna's olecranon process). (See Fig. 11.6.)

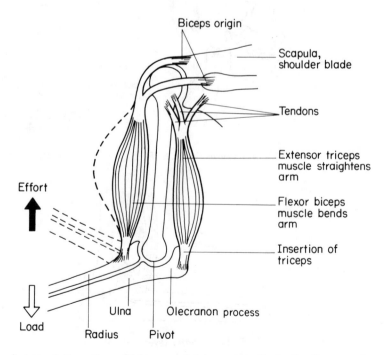

Figure 11.6 Antagonistic action of biceps and triceps muscles in the forearm. In examinations you will probably be asked to 'label and explain this figure'.

192

Arm straightening is an example of the forearm acting as a first-order lever:

LOAD ——→ PIVOT ——→ EFFORT

Arm bending is an example of the forearm acting as a third-order lever:

LOAD ——→ EFFORT ——→ PIVOT

11.2 Practical Work

(a) The Mammal Skeleton

The bones composing the following parts of a rabbit skeleton should be examined and diagrams drawn.
  (i) vertebral column;
 (ii) skull;
(iii) limbs and limb girdles.

(b) Muscle Tissue

Prepared microscope slides or 35 mm transparencies of skeletal striped muscle, smooth unstriped muscle and cardiac heart muscle should be examined.

(c) Energy Release in Skeletal Muscle

1. Arrange a strand of fresh butcher's meat on a microscope slide placed over graph paper.
2. Add a few drops of 1% glucose solution to the meat muscle fibres and observe any change in length.
3. Add a few drops of prepared ATP (adenosine triphosphate) solution and again observe any change in length of the meat muscle fibres.

Water and glucose alone cannot cause muscle fibre contraction, but ATP solution alone causes the necessary contraction of muscle.

(d) Flowering Plant Support

1. *Turgidity* of potato, rhubarb and dandelion stem tissues is demonstrated by immersion in plain water and in concentrated salt solution.
2. *Mechanical supporting tissues* are observed in prepared transverse sections of primary (unthickened) stems of buttercup and secondary-thickened stems of lime.

11.3 Examination Work

(a) Multiple-choice Objective Questions

1. Which of the following is an example of voluntary striped muscle?
    (a) eye ciliary muscle

(b) gut smooth muscle
(c) artery wall muscle
(d) arm biceps muscle

2. Which of the following is composed of contractile tissue?
   (a) skeleton bone
   (b) muscle tendon
   (c) bone ligaments
   (d) skeletal muscle

Questions 3 to 10 refer to the accompanying diagram showing part of a human arm.

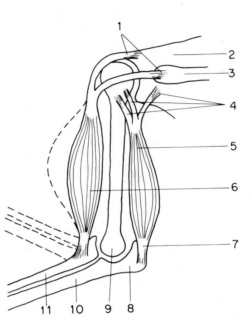

3. Bone labelled 11 in the diagram is the:
   (a) scapula
   (b) radius
   (c) ulna
   (d) humerus

4. The ball-and-socket joint of the forearm is between the:
   (a) radius and ulna
   (b) radius and scapula
   (c) ulna and humerus
   (d) scapula and humerus

5. The joint located at part labelled 9, the elbow, is called:
   (a) ball-and-socket joint
   (b) hinge joint
   (c) immovable suture
   (d) gliding joint

6. The part labelled 3 is a portion of the
   (a) collar bone
   (b) shoulder blade
   (c) rib bone
   (d) breast bone

7. The labelled part showing the point of insertion of a muscle is:
   (a) 1
   (b) 4
   (c) 6
   (d) 7

8. The muscle labelled 5 in the diagram is called the:
   (a) biceps
   (b) triceps
   (c) quadriceps
   (d) brachialis

9. Muscles labelled 5 and 6 in the diagram are examples of a pair of muscles called:
   (a) flexor muscles
   (b) extensor muscles
   (c) antagonistic muscles
   (d) smooth muscles

10. Which of the following is the direct cause of contraction of muscle labelled 6?
    (a) nerve impulse
    (b) acetyl choline
    (c) adenosine triphosphate (ATP)
    (d) glucose

11. Two bones are connected together at a joint by:
    (a) synovial membranes
    (b) ligaments
    (c) cartilages
    (d) tendons

12. The axial skeleton consists of the:
    (a) skull and vertebral column
    (b) limb girdles
    (c) limb bones
    (d) limb bones and girdles

13. Mammal skeletal muscles are attached to bones by:
    (a) tendons
    (b) cartilages
    (c) synovial membranes
    (d) joint capsules

14. The bone whose opposite ends form parts of the hip and knee joints is called the:
    (a) humerus
    (b) femur
    (c) pelvis
    (d) tibia

15. Movable joints contain a viscous fluid in the joint cavity called:
    (a) serum
    (b) tissue fluid
    (c) synovial fluid
    (d) blood plasma

16. Cartilage and bone are skeletal tissues found in:
    (a) earthworms
    (b) insects
    (c) mammals
    (d) molluscs

17. Which of the following is the contractile protein component of vertebrate muscle?
    (a) acto-myosin
    (b) myoglobin
    (c) glycogen
    (d) ATP (adenosine triphosphate)

18. The part of a movable joint which secretes the fluid in the joint cavity is the:
    (a) cartilage surface
    (b) joint ligaments
    (c) synovial membrane
    (d) joint capsule

19. The concave part of the mammal hip joint is called the:
    (a) glenoid cavity
    (b) acetabulum
    (c) femur head
    (d) humerus head

20. The heart and lungs are protected by the:
    (a) ribs
    (b) sternum
    (c) thoracic cage
    (d) thoracic vertebrae

(b) Structured Questions

Question 11.1

The accompanying illustration is a diagrammatic section through the pelvic girdle and the femur at the hip.

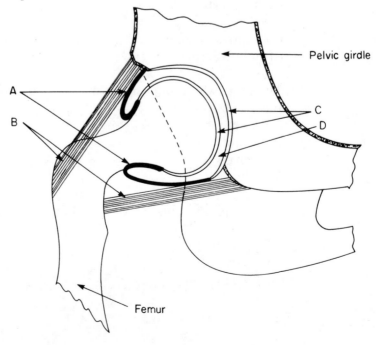

196

(i) Name the parts A, B, C and D.

(ii) What are the functions of C and D?

(iii) What type of joint is represented in the diagram?

(iv) Name TWO other different types of joint in the mammalian skeleton and briefly describe their function in relation to where they are situated.

(v) How is the structure of bone in the femur adapted to support the mass of a human?

(11 marks)

(NISEC)

* Question 11.2

Up to 40% of the body mass of a mammal is muscle tissue.

(a) What is a tissue? **(3 marks)**

(b) Name a muscle attached to the skeleton of a mammal. Explain, with the help of a diagram, how it is attached to the skeleton and describe the function it performs. **(8 marks)**

(c) Muscles surround the cavities of the alimentary canal (gut).

(i) Describe the arrangement of these muscles in the gut wall. **(3 marks)**

(ii) Explain how the action of these muscles moves food along the gut. **(6 marks)**

(d) Suggest why muscle contains:

(i) glycogen **(3 marks)**

(ii) phosphates in the form of adenosine diphosphate (ADP) and adenosine triphosphate (ATP) **(3 marks)**

(iii) a dark red pigment called myoglobin which is similar to haemoglobin **(2 marks)**

(e) The breast meat of a domestic chicken is white. A wild goose has dark-coloured breast meat. Suggest an explanation for this difference. **(2 marks)**

(AEB, 1981)

(c) Free-response-type Questions

(i) *Short-answer Questions*

1. Describe two features of bone tissue. Describe two features of cartilage. What type of synovial joint is present at the head of a femur, and the end of the femur forming the knee? **(6 marks)**

(AEB, 1981)

2. State the biological function or importance of:
(i) synovial fluid, (ii) ATP. **(2 marks)**

(SUJB)

3. State two functions of an animal skeleton apart from support. **(2 marks)**

(SEB)

(ii) Long-answer Questions

* Question 11.3

(a) List *four* functions of the skeleton of a mammal. **(2 marks)**

(b) Show how the structure of a synovial joint is related to its function of allowing movement to occur. Use a diagram in your answer. **(9 marks)**

(c) List the types of joint found in the skeleton. Give an example of the location of each type you mention. **(4 marks)**

(SUJB)

Answer to Q11.3

This is a common question on the mammal skeleton.

(a) Skeleton functions: 1, support; 2, movement; 3, protection of soft organs; 4, blood cell formation in red bone marrow.

(b) A synovial joint is a *movable* joint, lined partly by a *synovial membrane* secreting *synovial fluid*, bathing the joint cavity and lubricating the hyaline *cartilage* surfaces of two or more bones. Ligaments connect bones together and form a joint capsule which allows free movement of the joint in different planes depending on the type of joint (hinge, ball-and-socket). The part (a) of Fig. 11.3 showing a typical ball-and-socket joint is needed here.

(c) Joints in the skeleton are:
  (i) *fixed joints*, or sutures between skull bones
  (ii) *slightly movable joints* as between vertebrae
  (iii) *freely movable* (synovial) joints between limb bones — hinge, ball and socket.

✳ Question 11.4

Describe the movement of a mammalian limb, referring in your answer to the functions of muscles, bones and joints. **(25 marks)**

(L)

Answer to Q11.4

The description of mammal limb movement can be given by reference to human forearm movement. This answer requires a labelled diagram showing Fig. 11.6.

Bones involved include: scapula bone of shoulder girdle, humerus, radius and ulna.

Muscles involved: biceps, flexor with two origins, and triceps, extensor with three origins.

Joints: two joints, shoulder and elbow
  (i) between humerus and scapula and the glenoid cavity — ball-and-socket joint;
  (ii) between humerus and the radius and ulna, forming a hinge joint.

Movement

1. *Raising the forearm*, or bending the arm, in which the forearm acts as a load; involves contraction of biceps acting as the effort, resulting in movement across pivot of elbow joint. This is a third-order lever:

LOAD	→	EFFORT	→	PIVOT
forearm		biceps		elbow
		muscle		joint

2. *Straightening the arm*, or lowering the forearm, involves contraction of the triceps acting as the effort, causing movement across the elbow pivot. This is a first-order lever:

$$\text{LOAD} \longrightarrow \text{PIVOT} \longrightarrow \text{EFFORT}$$
$$\text{forearm} \qquad \text{elbow} \qquad \text{triceps}$$

* **Question 11.5**

(a) Make a labelled diagram to show the structure of the stem of a named herbaceous flowering plant as seen in transverse section. **(9 marks)**

(b) Explain how a herbaceous plant normally stays erect. **(5 marks)**

(c) (i) What internal changes cause wilting? **(2 marks)**

    (ii) What environmental changes bring this about? **(4 marks)**

(d) For each of the following organisms, name or briefly describe the material or skeletal system that gives it shape or rigidity: (i) a tree; (ii) a mammal; (iii) an adult insect; (iv) an earthworm; (v) a mould aerial hypha. **(5 marks)**

(OLE)

Answer to Q11.5

This is a *mixed* question, concerned with *support* in plants and animals. It requires a knowledge of *supporting tissues* and *water relationships* in plants, discussed in Chapters 3 and 6, with a knowledge of animal support as given in this chapter.

(a) Figure 3.2 needed here.

(b) *Herbaceous plants* stay erect by means of:
    (i) cell turgidity;
    (ii) cellulose walls;
    (iii) mechanical supportive tissue, cells with *thickened* cell walls, collenchyma and sclerenchyma;
    (iv) vascular bundles composed of xylem and phloem, arranged centrally in roots and cylindrically in stems, thus *reinforcing* plant organ support;
    (v) *climbing* herbaceous plants gain support by twining stems and tendrils.

(c) (i) *Wilting* is caused by the loss of water from cells by evaporation or transpiration being greater than water intake by roots. Cells lose their turgidity. Wilting also occurs in certain *diseased* plants.
    (ii) Environmental changes which cause wilting include those that affect the transpiration rate (Chapter 6): dry air; high temperature; wind; water shortage.

(d) Rigidity or shape is maintained in the following:
    (i) a tree by *secondary thickening* or growth tissues, mainly secondary xylem or wood;
    (ii) a mammal by its internal bony *endoskeleton* and by skeletal muscles;
    (iii) an adult insect by its *exoskeleton* or chitinous cuticle;
    (iv) an earthworm by its *hydrostatic skeleton* due mainly to rigidity of tissues, muscle contraction and internal fluid pressure;
    (v) mould aerial hyphae have cell walls composed of stiff fungal cellulose and can be stiffened with *chitin*.

11.4 Self-test Answers to Objective and Structured Questions

Answers to Multiple-choice Objective Questions

1. d   2. d   3. b   4. d   5. b   6. b   7. d   8. b   9. c   10. c   11. b   12. a
13. a   14. b   15. c   16. c   17. a   18. c   19. b   20. c

Answer to Structured Question 11.1

(i) Part A − synovial membrane *or* joint capsule
     Part B − joint ligaments
     Part C − hyaline cartilage
     Part D − synovial fluid

(ii) Functions of C and D: provision of *friction-free* surface for smooth joint cartilage surface *lubricated* by synovial fluid.

(iii) Type of joint is ball-and-socket.

(iv) Hinge joint in the knee allowing bending of leg. Immovable joint between the bones of the skull cranium.

(v) Femur bone is adapted for support in being composed of hard compact skeletal material in the bone shaft, and by virtue of the fact that the bone shaft is a strong hollow *cylinder*, giving greater strength than a dense *rod*.

Answer to Structured Question 11.2

(a) A tissue is a group of cells with a similar structure specialised to perform a similar function, for example *muscle*.

(b) A mammalian skeleton muscle is the forearm *triceps* shown in Fig. 11.6. Triceps is inserted or attached to ulna bone at the olecranon process by ligaments. It also has *three* points of origin via ligaments to the humerus head and scapula. Its function is to straighten the arm. This occurs when the triceps muscle *contracts* and the biceps muscle *relaxes*.

(c) (i) *Gut wall* muscles are smooth involuntary muscles arranged in two layers, *outer* longitudinal and *inner* circular muscle layer (see Fig. 11.1).
     (ii) Food is moved along the gut by sequential contraction of gut wall circular muscles resulting in waves of muscle contraction pushing food along the gut.

(d) (i) Glycogen is a carbohydrate store readily changed into glucose by enzyme action.
     (ii) Adenosine triphosphate (ATP) provides the energy for muscles to contract and produce ADP and phosphate.

$$\text{ATP} \longrightarrow \text{ADP} + \text{phosphate} + \text{energy}$$

     (iii) Myoglobin is a muscle protein that can combine with oxygen to form oxymyoglobin.

$$\text{Myoglobin} + \text{oxygen} \longrightarrow \text{oxymyoglobin}$$

(e) The dark-red colour of the breast muscle of a wild goose is partly due to myoglobin; such muscle is used for prolonged strenuous exercise as in flight, requiring aerobic respiration. Domestic chicken breast muscle is white because of the absence of myoglobin; these muscles are not used for the strenuous activity of flight since the chicken is almost flightless.

Theme III

Development of Organisms and Continuity of Life

12 Reproduction

12.1 Theoretical Work Summary

Life, which has been maintained by *nutrition*, *respiration* and *transport*, and which has been protected through *irritability* and *response*, is allowed to continue in new living *body cells* by *growth* and in *individuals* by *reproduction*.

(a) Asexual and Sexual Reproduction

Reproduction is the continuation of life processes in new organisms. There are two types, sexual and asexual (see Table 12.1).

Table 12.1 A comparison of sexual and asexual reproduction

Sexual	Asexual
Two parents	Single parent
Male and female gametes produced by gametogenesis	No gamete production
Diploid zygote formed by fertilisation	No zygote formation
Meiosis essential for gamete formation	Mitosis essential for spore formation and cell division
Variation in offspring, with hybrid vigour	Offspring identical to parent; i.e. a clone
Not rapid	Rapid in favourable conditions
Population numbers increase slowly	Population numbers increase rapidly
Occurs among all living organisms	Mainly among plants and simpler invertebrate animals
By sexual organs, in cones, flowers, testes and ovaries	By cell division, binary fission, fragmentation, budding, spores, vegetative reproduction and artificial propagation

(b) Asexual Reproduction

This produces organisms or cells that are *identical* to one another and are known as *clones*. Figure 12.1 shows the method of asexual reproduction in *Mucor.*

(c) Vegetative Reproduction and Perennation

These are examples of methods used to produce clones.
1. *Vegetative reproduction (propagation)* is an asexual method in which specialised *multicellular organs* (bulbs, corms, tubers, rhizomes or runners, Fig. 12.2) form and become detached from the parent and give rise to new individual plants.
2. *Perennation* by *biennial* and *perennial* herbs and by shrubs and trees, is a means of survival of adverse conditions in winter. It also occurs by means of *food*

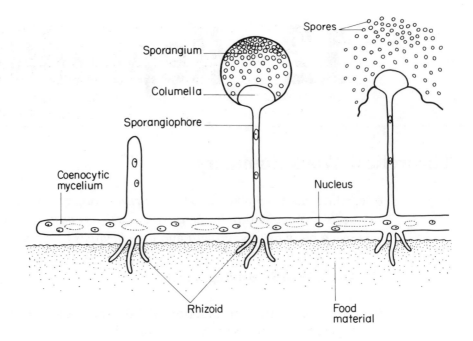

Figure 12.1 Asexual reproduction in *Mucor* sp.

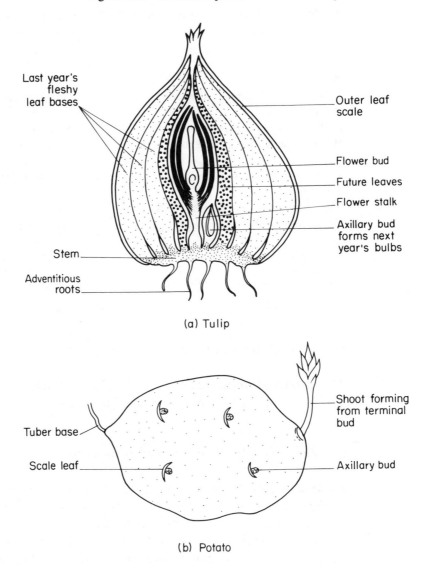

(a) Tulip

(b) Potato

Figure 12.2 Vegetative reproduction organs: (a) bulb, (b) tuber.

storage in *seeds*, swollen underground *vegetative organs*, and dieback of aerial parts by *leaf fall* (see Fig. 12.3).

3. *Artificial propagation* is a form of vegetative propagation by budding, layering, grafting and stem or leaf cuttings rooted by plant hormones (*auxins*) (Fig. 12.4).

4. *Tissue culture* is the growth of pieces of plant or animal tissue in a sterile solution of nutrients supplied with oxygen and maintained at 38°C.

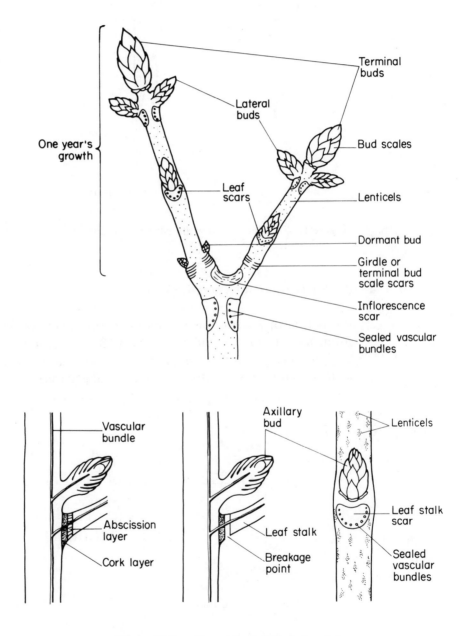

Figure 12.3　Horse chestnut twig in winter.

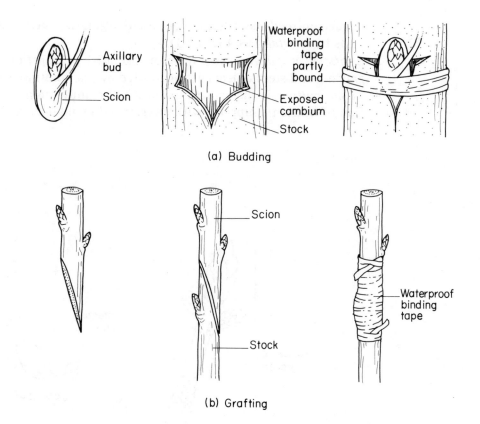

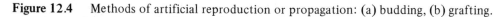

(a) Budding

(b) Grafting

Figure 12.4   Methods of artificial reproduction or propagation: (a) budding, (b) grafting.

✻ (d) Alternation of Generations

Alternation of generations is seen in the *life cycle* of parasitic animals, e.g. tapeworm, and in *mosses* and *fern* plants. *Two* reproductive forms occur, a *sexual gamete* producer and an *asexual spore* producer, in different *generations*; the sexual generation is followed by an asexual generation.

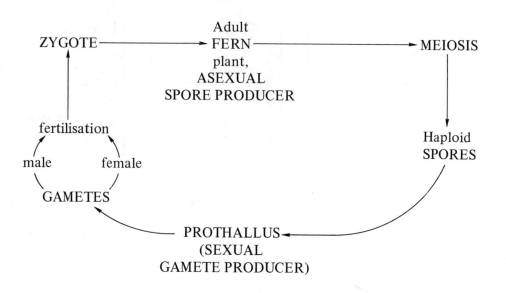

(e) Sexual Reproduction in Seedless Plants

Seeds are produced sexually in *flowering* and *coniferous* plants. All other plants do not produce seeds sexually.

1. *Zygospores* are formed sexually by fusion of gametes formed from whole individuals in algae (e.g. *Spirogyra*) and fungi (e.g. *Mucor*). Zygospores have a thick resistant protective wall: see Figs 12.5(a) and (b).
2. *Oospores* are formed sexually by fertilisation of the female gamete (*oosphere*) by male gametes (*spermatozoids*) both produced by the fern *prothallus*, the haploid generation of the life cycle: see Fig. 12.5(c).

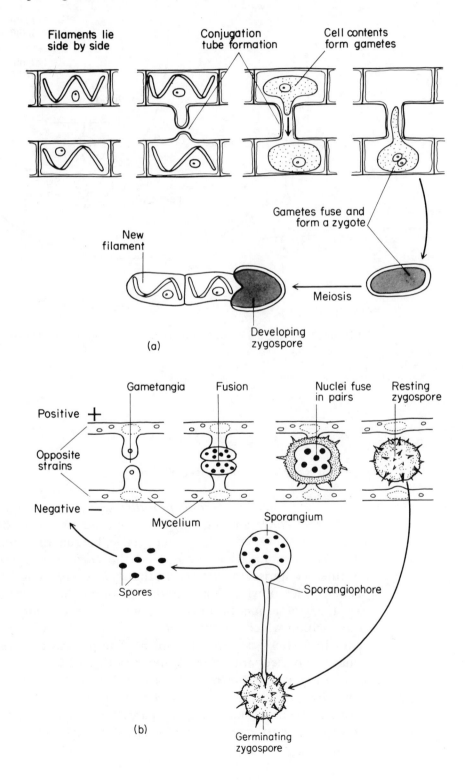

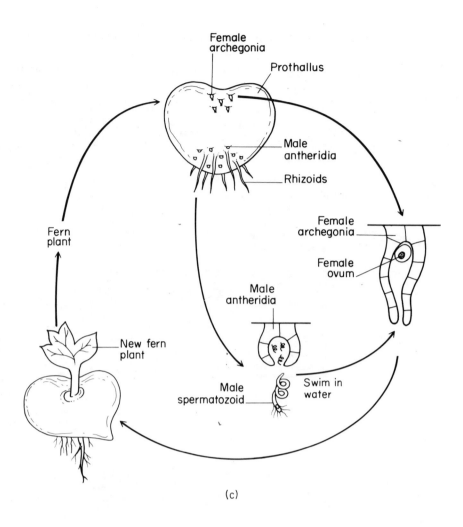

Female archegonia

Prothallus

Male antheridia

Rhizoids

Female archegonia

Female ovum

Male antheridia

Fern plant

New fern plant

Male spermatozoid

Swim in water

(c)

Figure 12.5   Sexual reproduction in (a) *Spirogyra* sp., (b) *Mucor* sp., (c) *Dryopteris* sp.

(f) Sexual Reproduction in Seed Plants

1. *Flowers* are usually the *hermaphrodite* sexual reproductive units of *flowering* plants, and *cones* are the *unisexual* reproductive units of *conifers*. A typical angiosperm flower consists of a receptacle, bearing the remaining parts, namely: the *calyx* composed of *sepals*, and the *corolla* composed of *petals*, which together comprise the *perianth*; the *androecium*, consisting of a number of *stamens*, each of which has an *anther* (forming the pollen), borne on a *filament*: and the *gynoecium* or pistil, comprising one or more *carpels* containing an *ovary* within which are the *ovules*; the carpel often bears a *style* which terminates in a sticky surface, the *stigma*. The process of sexual reproduction in a typical dicotyledonous plant is shown in Fig. 12.6.

2. *Pollination* can occur by *self-* or *cross-pollination*, and is brought about by wind, by water or by an insect or other animal transferring pollen from an anther stigma. Table 12.2 compares the characteristics of insect-pollinated flowers (e.g. buttercup and sweet-scented flowers) with those of wind-pollinated species (e.g. grasses).

Table 12.2 Features of insect- and wind-pollinated flowers

Insect pollination	Wind pollination
Coloured, scented, large flowers with nectary	Small flowers without scent, nectar, petals or sepals
Pollen dense and sticky. Stamens specially arranged within the flower	Pollen light and smooth surfaced, produced on large pendulous stamens which project outside the flower
Stigma small and sticky inside flower	Stigma large and feathery projecting outside flower

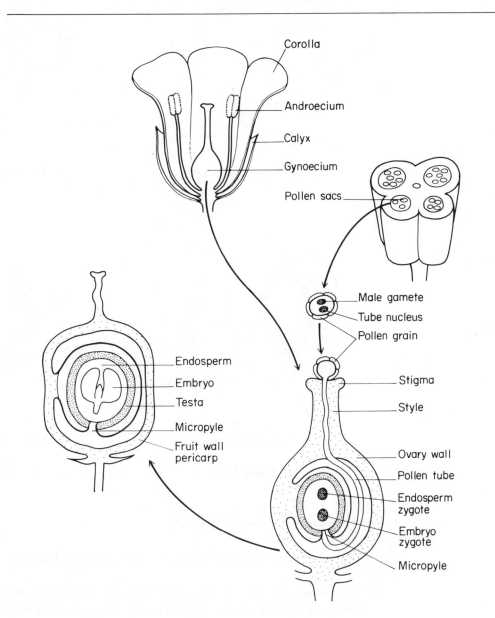

Figure 12.6 Pollination and fertilisation of a dicotyledonous flowering plant. In examinations you will probably be asked to 'label and explain this figure'.

3. *Fertilisation* is the fusion of haploid male and female *gametes* to form a diploid *zygote* from which an *embryo* develops contained within a *seed* surrounded by a *fruit*. *Internal fertilisation* occurs in flowering and coniferous plants, *external* fertilisation in mosses, liverworts and ferns.

4. *Fruits* are structures developing from the ovary wall or *pericarp* and serve to protect and disperse the seeds. They are of two main types, *dry* and *succulent* (see Fig. 12.7). The first category can be further subdivided into *dehiscent*, which burst to release the seeds (e.g. the legume of pea, the follicle of wall-flower and the capsule of poppy), and *indehiscent*, which do not burst (e.g. the achene of buttercup, the nut of hazel, and the caryopsis of wheat). Examples of succulent fruits are the drupe (as in plum) and the berry (as in tomato). *Dispersal* of fruits and seeds is by animals, wind, water or explosion.

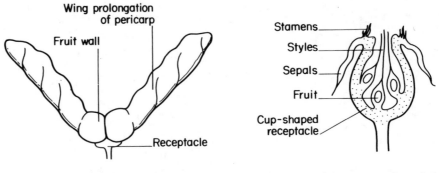

(a) Sycamore — dry indehiscent

(b) Rose hip — false fruit

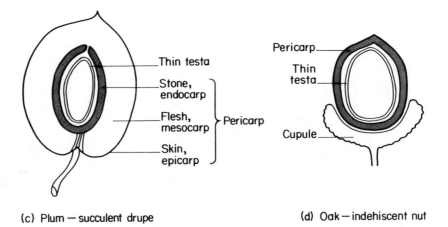

(c) Plum — succulent drupe

(d) Oak — indehiscent nut

Figure 12.7 Examples of dry and succulent fruits.

(g) Sexual Reproduction in Mammals

This may be described with reference to human beings (see Fig. 12.8).
1. *Gametogenesis*
    (a) The male gametes are the *spermatozoa*, which are produced in the *semi-niferous tubules* of the *testes* and are stored temporarily in the *epididymis*. *Seminal fluid* is a nutritive fluid from the seminal vesicles, Cowper's and prostate glands. *Ejaculation* of semen is via the *vas deferens* and *urethra*.
    (b) The female gametes, the *ova*, are produced in ovaries in the *ovulation cycle* or *menstruation cycle*, which in humans is of one month's duration. Ovulation involves *Graafian follicle* formation, release of the ovum accompanied by hormones and *oestrus*, and formation of *corpus luteum*. Meanwhile, the uterus lining is prepared for *implantation* of the embryo, or is renewed after bleeding in menstruation.

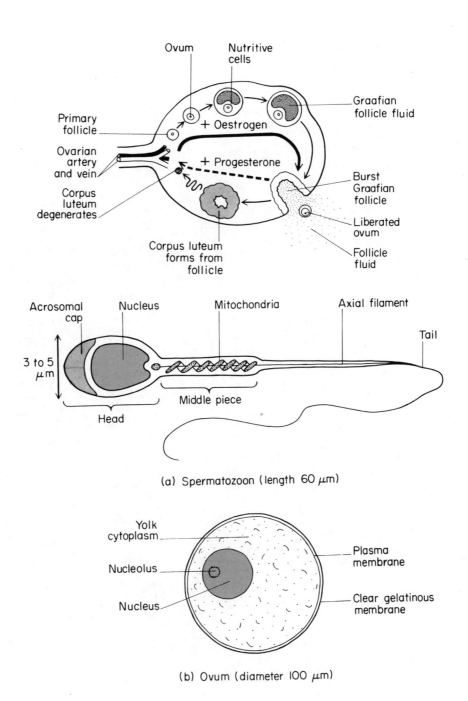

(a) Spermatozoon (length 60 μm)

(b) Ovum (diameter 100 μm)

Figure 12.8 Top: stages in the formation of an ovum in a human female. Bottom: human gametes.

2. *Fertilisation* is internal by ejaculation of semen into *vagina* by introduction of erect *penis* in sexual intercourse (*copulation, insemination* or *conjugation*). Zygote formation occurs in *fallopian tubes* (oviducts). (See Fig. 12.9.)
3. Development of the *embryo* (which is called a *foetus* after 4–6 weeks): *implantation* of embryo in the uterus lining, followed by *mitosis* and formation of tissues of embryo and embryonic membranes:
   (i) *Amnion* with *amniotic fluid* is a *buoyant* supporting medium providing *protection* for developing tissues and *steady* conditions; as a *lubricant* it allows free movement of the foetus.
   (ii) The *placenta* is formed between embryo *chorion* membrane and uterus lining. Small *villi* with blood capillaries bathed by mother's blood provide

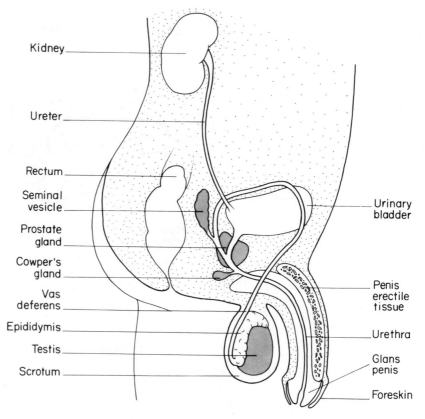

(a)

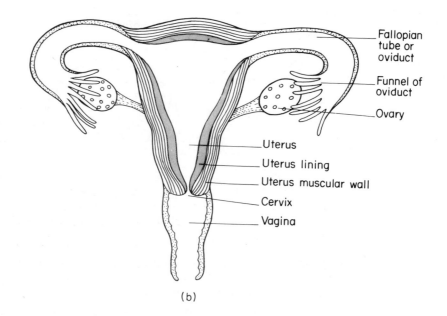

(b)

Figure 12.9 Human reproductive systems: (a) male, (b) female. In examinations you will probably be asked to 'label and explain this figure'.

nutrients and remove excretory waste via blood vessels of the *umbilical cord*. See Fig. 12.10.

4. *Birth* or *parturition* occurs after 9 months' *gestation*, the period between fertilisation and birth. *Oxytocin*, a hormone secreted by the pituitary, causes *contraction* of the uterus in labour and expulsion of the foetus via the vagina followed by the *afterbirth* (placenta and embryonic membranes) together with amniotic fluid *waters*.

5. *Lactation* or formation of milk in the breast. The lactating mother needs extra water and nutrients in her diet to form about 850 cm³ of milk daily.

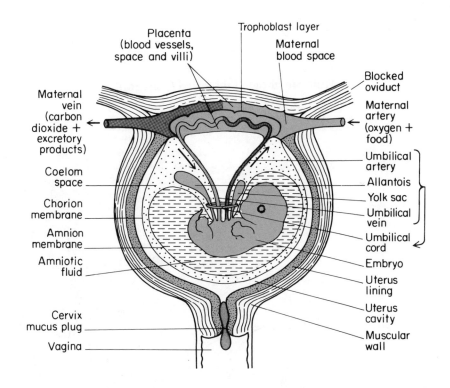

Figure 12.10 Human embryo in the uterus. In examinations you will probably be asked to 'label and explain this figure'.

12.2 Practical Work

(a) Plants

The main practical work involves

1. *Spore-producing plants* (fungi, mosses and ferns): the examination of the spore-producing organs by means of hand lenses.
2. *Vegetative reproduction* in flowering plants by examination of runners, rhizomes, stem tubers and bulbs.
3. *Winter twigs* of deciduous trees as examples of *perennation*.
4. *Flowers* of flowering dicotyledonous plants are examined by cutting vertical sections with a sharp razor and determining a simple floral formula for the number of flower parts using the symbols K, calyx; C, corolla; A, androecium; G, gynoecium.

    Simple keys can be used as given in wild-flower handbooks, using flower *colour*, for example, as a means of identification.
5. *Fruits* of different wild flowering plants are examined and classified according to structure and methods of dispersal.

(b) Animals

Mammal Reproduction

1. Prepared slides or 35 mm transparencies of *testis*, showing seminiferous tubules, and *ovary*, for developing graafian follicles, should be examined.
2. Preserved and mounted specimens of rat or rabbit embryos with embryonic membranes should be examined.

12.3 Examination Work

(a) Multiple-choice Objective Questions

Questions 1 to 10 concern the accompanying diagram.

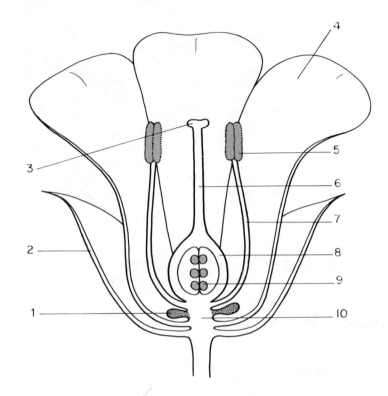

In examinations you will probably be asked to 'label and explain this figure'.

1. The vertical section of the flower shown in the diagram indicates it belongs to the group of plants called:
   (a) angiosperms
   (b) pteridophytes
   (c) gymnosperms
   (d) bryophytes

2. Pollen grains are formed at the part labelled:
   (a) 1
   (b) 3
   (c) 5
   (d) 9

3. Fertilisation takes place in the part labelled:
   (a) 7
   (b) 8
   (c) 9
   (d) 10

4. The two parts that continue to develop after fertilisation are:
   (a) 1 and 2
   (b) 4 and 5
   (c) 6 and 7
   (d) 8 and 9

5. The part producing a sugary fluid is:
   (a) 1
   (b) 3
   (c) 4
   (d) 5

6. The male gametes are formed in the part labelled:
   (a) 1
   (b) 5
   (c) 7
   (d) 9

7. Which labelled part develops into the seed after fertilisation?
   (a) 3
   (b) 5
   (c) 8
   (d) 9

8. The part labelled 2 is called the:
   (a) petal
   (b) sepal
   (c) stamen
   (d) floral axis

9. Which two labelled parts represent the ovary and ovule?
   (a) 2 and 4
   (b) 3 and 6
   (c) 5 and 7
   (d) 8 and 9

10. The carpel in the diagram comprises three parts:
    (a) 1, 2 and 4
    (b) 2, 4 and 8
    (c) 3, 6 and 8
    (d) 4, 5 and 7

11. The process of fusion of male and female gametes in mammals is called:
    (a) pollination
    (b) copulation
    (c) fertilisation
    (d) insemination

12. The mammalian placenta:
    (a) surrounds and protects the embryo
    (b) connects the maternal and foetal blood systems
    (c) is a store of food, mainly protein and lipids
    (d) is a store for excretory waste products

13. The membrane that surrounds and encloses a developing mammal foetus is called the:
    (a) placenta
    (b) amnion
    (c) umbilicus
    (d) uterus

14. Following fertilisation the fertilised mammalian egg will be found embedded in the:
    (a) ovary
    (b) uterus

(c) vagina

(d) vulva

15. The tube in a male mammal along which gametes travel from the testis to the urethra is called the:
    (a) ureter
    (b) vas deferens
    (c) penis
    (d) fallopian tube

16. Spermatogenesis occurs in the male mammal in the:
    (a) testis
    (b) Cowper's gland
    (c) prostate gland
    (d) seminal vesicles

17. Which one of the following does *not* pass across the placenta and the foetus?
    (a) red blood cells
    (b) blood plasma
    (c) antibodies
    (d) oxygen

18. Which of the following glands or structures is responsible for the development of secondary sexual characteristics in the female human being?
    (a) placenta
    (b) ovary
    (c) thyroid
    (d) uterus

19. A plant that lives for many years is called:
    (a) a perennial
    (b) a biennial
    (c) an annual
    (d) an ephemeral

20. Which of the following causes the uterus to contract and expel the foetus by hormonal secretion?
    (a) ovary
    (b) placenta
    (c) pituitary
    (d) thyroid

(b) Structured Questions

* **Question 12.1**

After fertilisation in humans the resulting zygote develops into a ball of cells which becomes embedded in the lining of the uterus. Subsequently, the embryo and placenta develop, together with the membranes which enclose the embryo.

(a) (i) What do you understand by the term *fertilisation*? **(2 marks)**

(ii) Where in the body does fertilisation which would lead to pregnancy usually take place? **(1 mark)**

(iii) What is a zygote? **(2 marks)**

(iv) Explain how the placenta functions both as an exchange surface for dissolved gases and as an excretory organ for the embryo. **(6 marks)**

(b) During pregnancy the embryo becomes surrounded by a fluid (amniotic fluid) which is enclosed with the embryo inside the membranes formed after implantation. As pregnancy progresses, some cells of the embryo appear suspended in the fluid. The diagram shows the main stages of amniocentesis, which allows the withdrawal and examination of amniotic fluid.

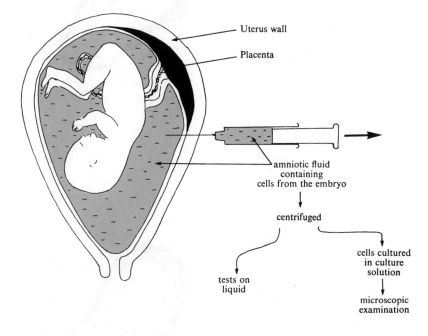

(i) State *two* normal functions of amniotic fluid. **(2 marks)**

(ii) What is the purpose of centrifuging the extract obtained by amniocentesis? **(2 marks)**

(iii) How may cells from the fluid be used for determining the sex of an unborn child? **(4 marks)**

(iv) Suggest *two* ways in which cells from the embryo could find their way into the fluid. **(2 marks)**

(v) Suggest *three* features essential in a culture medium which is used to keep human cells alive and active outside the body. **(3 marks)**

(c) The simplified diagrams show two stages in the birth of a baby.
(i) Name X. **(1 mark)**

(ii) State *two* major events which have occurred between the two stages other than the emergence of the head. **(2 marks)**

(iii) Stage 2 shows the child's head emerging. Briefly describe the events which follow immediately afterwards. **(3 marks)**

(AEB, 1983)

Question 12.2

The diagrams represent fruits which are dispersed by various methods. Complete the table, placing letters from the diagrams in the correct box:

Animal dispersal	Wind dispersal

Question 12.1

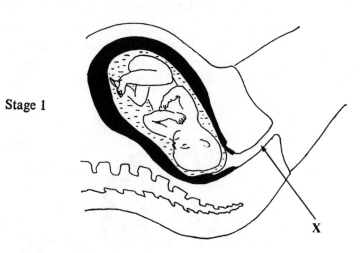

Stage 1

X

Stage 2

Question 12.2

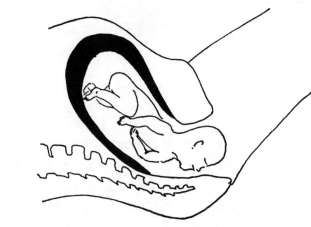

N

S

T

P

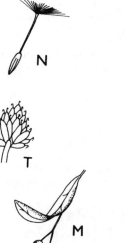

M

(i) The diagram that follows represents the flower of a grass plant. State two features, illustrated by the diagram, which aid the plant's method of pollination. **(2 marks)**

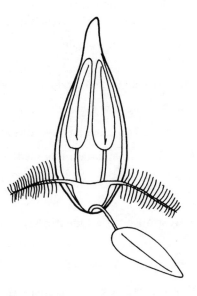

(ii) From which part of the flower does the seed develop? **(1 mark)**

(SEB)

✳ Question 12.3

(a) The diagram shows half a flower of the common gorse (*Ulex europaeus*), which is an insect-pollinated flower.

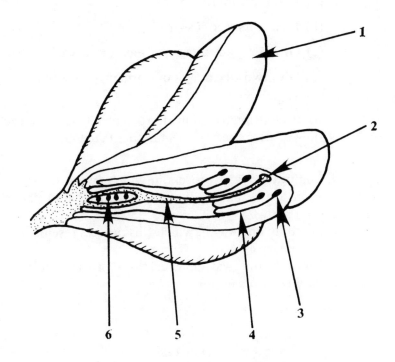

(i) *Name* the structures labelled 1 to 6 on the diagram. **(6 marks)**

(ii) Explain what you understand by the term pollination. **(2 marks)**

(iii) State *two* features shown in the diagram which indicate that this flower is insect-pollinated. **(2 marks)**

(iv) State *two other* features you would expect an insect-pollinated flower to possess.

(2 marks)

(v) Describe the events following pollination which lead to the formation of a seed.

(10 marks)

(b) The seeds of the common gorse are contained in pods. Some time after their formation these pods can be seen open and empty. Their walls are then very dry.

(i) What is the biological name given to the pod? **(1 mark)**

(ii) Describe how the seeds of the common gorse are dispersed. **(3 marks)**

(iii) Name two other methods of seed dispersal. **(2 marks)**

(iv) Why is it important that seeds are dispersed? **(2 marks)**

(AEB, 1982)

(c) Free-response-type Questions

(i) *Short-answer Questions*

1. (a) Name a species of organism that produces both zygotes and asexual spores in its life cycle. **(4 marks)**

   (b) What is a zygote? **(3 marks)**

   (c) What is an asexual spore? **(2 marks)**

   (d) Give three advantages to an organism of being able to produce both zygotes and asexual spores in its life cycle. **(3 marks)**

   (OLE)

2. (a) Name a flowering plant.

   (b) How is its seed dispersed? **(1 mark)**

   (c) Give three advantages of seed dispersal. **(3 marks)**

   (d) Give two advantages of dormancy in seeds. **(2 marks)**

   (OLE)

3. What contribution is made to reproduction by each of the following parts of a flower: (i) anther, (ii) nectary, (iii) stigma? **(3 marks)**

   (SUJB)

4. Conjugation is a process in which two ordinary bacterial cells come together, form a tube linking the two cells, exchange part of their genetic material and then part.

   (a) State one way in which this process resembles sexual reproduction in higher organisms.

   (b) State one way in which this process differs from sexual reproduction in higher organisms. **(2 marks)**

   (AEB, 1982)

(ii) *Long-answer Questions*

✳ Question 12.4

(a) How does sexual reproduction differ from asexual reproduction? **(4 marks)**

(b) Describe how the mammalian embryo is: (i) protected and (ii) provided with nutrients.

(12 marks)

(L)

Answer to Q12.4

(a)

Sexual	Asexual
1. Two parents	1. One parent
2. Difference in offspring (except in identical twins)	2. Offspring identical
3. Male and female gametes required	3. Gametes not required
4. Slow process sometimes without increase	4. Rapid process with increase in numbers

(b) A diagram is required here, similar to Fig. 12.10, showing the embryo within the uterus and the embryonic membranes.
    (i) *Protection of the embryo* is provided by:
        1. the mother's *body* carrying the developing embryo internally;
        2. the strong muscular *uterus*;
        3. the *amnion* and the *amniotic* fluid provide further hydrostatic protection from mechanical damage, and a constant fluid environment.
    (ii) *Nourishment* is provided:
        1. initially by *implantation* in the *uterus lining* which feeds the embryo via its *glandular* secretion.
        2. *The placenta* is the main means of nourishment of the embryo and foetus. It is formed from the endometrium (uterus lining) of the mother and the *chorion* of the embryo. *Placental villi* contain blood capillaries, which connect with the embryo via blood vessels in the *umbilical cord*; these are bathed in maternal blood. The maternal blood provides: food, oxygen and antibodies, and removes urea (carbamide) and carbon dioxide.

✳ **Question 12.5**

    (a) Give a clearly labelled diagram of the reproductive system of a *named* female mammal.

(6 marks)

    (b) How does the female mammal protect and nourish her young from the zygote stage until birth?

(9 marks)

(SUJB)

Answer to Q12.5

(a) A large, well labelled diagram of Fig. 12.9(b) is needed here.

(b) This part of the question requires an answer identical to that given for question 12.4(b).

✳ Question 12.6

Describe the process of sexual reproduction in a *named*, insect-pollinated, flowering plant. In your answer refer to pollination, fertilisation, seed and fruit formation and dispersal.

(25 marks)

(L)

Answer to Q12.6

This is a common direct question from many examining boards. The key words give an indication as to how to set out the answer in *five* parts.

Insect-pollinated Flowering Plant

Sexual reproduction involves the specialised reproductive organ, the *flower*. The name of the insect-pollinated plant must be given, for example the primrose, and a vertical-section diagram must be drawn to its structure. (A *generalised* diagram should *not* be drawn.)

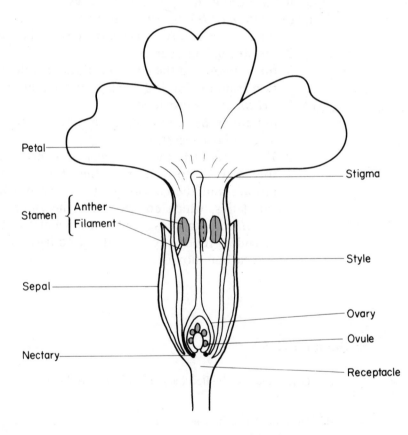

Pollination

Bees insert their proboscis, already covered in pollen from another primrose flower, into the corolla tube and the pollen attaches itself to the sticky stigma. The bee collects nectar from the nectaries and withdraws its proboscis, again covered in fresh pollen which will *cross-pollinate* another primrose flower.

Pin-eyed primroses have a long style and short stamens, as in the diagram, whereas *thrum-eyed* primroses have a short style and long stamens near the corolla tube opening. This difference in flower form aids cross-pollination. Differences in pollen grain size, and stigma structure are further modifications to ensure cross-pollination between different forms.

Fertilisation

A *pollen tube* grows out of the pollen grain and downwards through the *style*, feeding on its tissues. The pollen tube reaches the ovary and its *male gamete* fuses with the *female gamete* of the *ovule* to form the *zygote*.

Seed and Fruit Formation

The zygote nucleus divides by mitosis to form the seed *embryo* (the radicle, plumule and cotyledon). The seed is attached to the ovary wall by a placenta, which provides nutrients for the developing seed within the *fruit* or ripened ovary.

Seed Dispersal

The *ripe* fruit is a *dry capsule* that opens by means of five small openings, which allow the small dry seeds to escape when wind shakes the capsule; the seeds fall close to the parent plant.

12.4 Self-test Answers to Objective and Structured Questions

Answers to Multiple-choice Objective Questions

1. a 2. c 3. c 4. d 5. a 6. b 7. d 8. b 9. d 10. c 11. c 12. b 13. b 14. b 15. b 16. a 17. a 18. b 19. a 20. c

Answer to Structured Question 12.1

(a) (i) *Fertilisation* is the fusion of male and female gametes during sexual reproduction.
  (ii) Female *fallopian tubes.*
  (iii) A cell formed by fusion of two gametes — it is *diploid*.
  (iv) Exchange of these substances, dissolved gases and excretory products, occurs between blood capillaries within *villi* of the placenta which project into small blood-filled spaces provided from the mother's blood. Exchange is by diffusion: there is *no* mixing of foetal and mother's blood.

(b) (i) Provides steady external temperature in a fluid environment, acts as a buoyant and protective medium, and as a *lubricant*, and allows the foetus to move freely.
  (ii) Centrifuging separates the cells from the fluid.
  (iii) Photomicrographs of cell nuclei will show *XX* chromosomes in a female and *XY* chromosomes in a male. See page 232.
  (iv) Cells accumulate in the fluid, either as shed *skin epidermis* or as cells from the *embryonic membranes*, e.g. placenta.

(v) *Tissue* cultures need *warm sterile* microbe-free solutions containing *glucose*, *amino acids* and *mineral* ions, and dissolved *oxygen*, to keep the cells alive and active.

(c) (i) Vagina.

(ii) Bursting of the amnion and release of amniotic fluid 'waters'. Contraction of the uterus in labour.

(iii) Breathing commences, followed by expulsion of afterbirth (the placenta and embryonic membranes).

Answer to Structured Question 12.2

Animal dispersal	Wind dispersal
P T S	N M

(i) Method of pollination is by *wind* because of large hairy stigmas and large pendulous anthers.

(ii) Seed develops from the ovule within the ovary.

Answer to Structured Question 12.3

(a) (i) Labelled structures are: 1. petal, 2. stigma, 3. anther, 4. filament, 5. style, 6. ovule.

(ii) Pollination is the transfer of pollen from an anther to a stigma in flowering plants.

(iii) Insect pollination is indicated by large showy petals and small anthers.

(iv) Other features would include nectaries, producing a sugary solution; coloured and scented petals; pollen grains dense and sticky to cling to insects; short filaments.

(v) Pollen on stigma would produce pollen tube which would penetrate the style and enter the ovary. The male gamete nucleus in the pollen tube would fuse with the female gamete nucleus of the ovule to form a zygote. The zygote would undergo mitotic division and form the embryo (radicle, plumule and cotyledons) surrounded by or containing the endosperm of the seed. Meanwhile the ovary wall would develop into a fruit wall or pericarp.

(b) (i) Pod is called a legume.

(ii) Pods split lengthwise with explosive violence, scattering the seeds over considerable distances.

(iii) Seeds can be dispersed by animals or wind.

(iv) Dispersal is important to ensure survival of the species away from the parent plant and the shade that it cast, which would deprive the plant of light.

13 Genetics and Evolution

13.1 Theoretical Work Summary

Genetics is the study of *inheritance* or transmission of *genetic information* from generation to generation.

(a) The Nucleus

In eukaryotes this important organelle, which contains the genetic information, consists of the following:

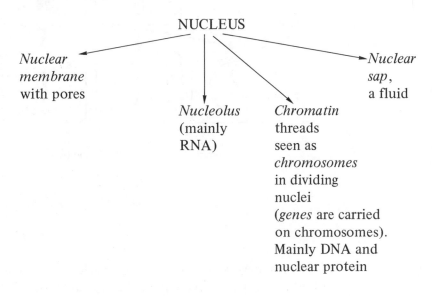

NUCLEUS

Nuclear membrane with pores

Nucleolus (mainly RNA)

Chromatin threads seen as *chromosomes* in dividing nuclei (*genes* are carried on chromosomes). Mainly DNA and nuclear protein

Nuclear sap, a fluid

(b) Chromosomes and Genes

Genes are units of genetic information or inherited material. *Chromosomes* are thread-like strands composed of nuclear protein and DNA, and carry the genes on positions called *loci*.

Replication is the important process by which the genetic material of the chromosomes is *duplicated* by forming *two* new molecules of DNA during the cell *interphase* or *growth period*, when the nucleus is in its *resting* stage.

Homologous chromosomes are *pairs* of chromosomes which separate in *meiosis* (Fig. 13.1). *Alleles* are the genes on opposite *loci* of homologous chromosomes.

The *chromosome number* is the number of chromosomes in normal body cells. This is constant for a species and is called the *diploid number*, $2n$ (in human beings $2n = 46$, and in garden pea $2n = 14$). The chromosomes can be seen and counted in photomicrographs.

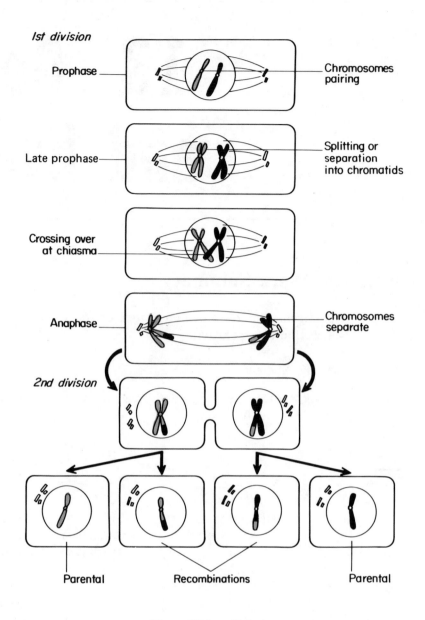

1st division

Prophase — Chromosomes pairing

Late prophase — Splitting or separation into chromatids

Crossing over at chiasma

Anaphase — Chromosomes separate

2nd division

Parental   Recombinations   Parental

Figure 13.1   Meiosis.

Chromosomes in normal body cell nuclei are of two main kinds:

(a) *Sex chromosomes.* One pair, usually called the *X* and *Y* chromosomes.

(b) *Autosomes.* More than one pair of chromosomes, and other than the sex chromosomes.

Genes are responsible for *coding* the synthesis of functional and structural proteins; consequently they are responsible for the external structure and internal physiology, i.e. all the characters shown by the organism in its *phenotype*, which interact with the environment. *Characters* or *traits* are structural or functional features controlled by *pairs* of genes, e.g. height, flower colour, blood group. The *genotype* is the set of genes that go to make up an organism.

(c) Meiosis (Reduction-Division)

Both mitosis (see Chapter 8) and meiosis are types of cellular and nuclear division. The processes are compared in Table 13.1.

Table 13.1   Comparison of mitosis and meiosis

	Mitosis	*Meiosis*
Occurrence	In body cells during *growth* and tissue *repair*	In reproductive cells to form *gametes*
Products	Two cells from one	Four cells from one
Chromosome number	Remains the same or is normal *diploid* (*two* sets of homologous chromosomes)	Is reduced by a *half* (*haploid*: *one* set of chromosomes)
Variation in chromosome structure	No variation	Variation occurs through *chromosome mutation* or *recombination*

(d) Variations

Variations are seen among *individuals* in a *population*. (See Section 14.1b.)

1. *Sexual reproduction* produces *variations* which are seen in the offspring's *phenotype* or external appearance.
2. *Asexual reproduction* does *not* produce any variation in offspring, e.g. as in vegetatively produced plants, or *identical* twins produced from one zygote having divided into two, each cell separating and forming two identical organisms (see Section 12.1b).
3. *Discontinuous variations* are *distinct*, e.g. flower colour, ability to taste a substance or to roll the tongue.
4. *Continuous variations* are *indistinct*, e.g. differences in height or weight.

Causes of Variation

Mutations or sudden changes in chromosomes or genes can cause variation. It is random and unpredictable, and can occur in meiosis or by effect of certain drugs and radiations.

Environmental changes, e.g. food, water, or light availability, affect growth, height or intelligence through preventing an organism reaching its full genetic potential.

(e) Inheritance or Heredity

1. *Genetics* is the study of inheritance of characters or traits by way of the *gene* material.
2. *Fertilisation* is the process of pairing of homologous chromosomes, provided by *haploid* gametes, to restore the species chromosome number in the *zygote*, continued by *mitosis* to form body (*somatic*) cells.
3. The mechanism of inheritance is called *Mendelism. Monohybrid* inheritance of a character such as flower colour in garden pea is due to the effect of *one pair* of genes on homologous chromosomes. *Dominant* characters or traits are seen in the progeny or offspring of *cross-fertilisation* between *pure-line* parents with contrasting characters, e.g. red and white flowers. The dominant character is red, and white is a *recessive* character. This *dominance* is seen in the $F1$ or first filial generation.
4. *Homozygous* individuals are formed from gametes having the identical genes or alleles, e.g. the gene for red flower colour, written as *RR* for *dominant* characters; *rr* is used for homozygous recessive zygotes.

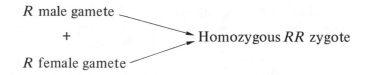

R male gamete

\+ $\longrightarrow$ Homozygous RR zygote

R female gamete

5. *Heterozygous* individuals are formed from gametes having contrasting genes, e.g. where the gene for red flower is R and white flower is r:

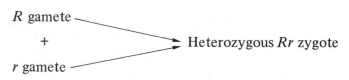

R gamete

\+ $\longrightarrow$ Heterozygous Rr zygote

r gamete

Since the dominant gene character is *red*, the individual heterozygote shows red flowers.

6. *Incomplete dominance* occurs when *neither* of the alleles or genes in the pair shows dominance. For example, when red-flowered snapdragon homozygous plants are crossed with white-flowered snapdragon homozygous plants, *pink* flowers are produced. (See also Fig. 13.2.)

(f) The Punnett Square

This is the chequerboard diagram (named after the geneticist R. C. Punnett) used to illustrate the formation of different zygotes and forecast the phenotype from the genotype. The female gametes are placed on the *right* side of the square and the male gametes on the *left* side.

Gametes of female homozygous for tallness

Gametes of male homozygous for shortness

	T	T
t	tT	tT
t	tT	tT

All $F1$ progeny are *heterozygous*, tT or Tt, and show the dominant character of tallness.

The consequences of self-fertilisation or inter-breeding of the $F1$ generation together are as follows:

Gametes of female heterozygous tall

Gametes of male heterozygous tall

	T	t
T	TT tall	Tt tall
t	tT tall	tt short

The progeny will be composed of:
25% homozygous tall, TT,
50% heterozygous tall, Tt
25% homozygous short, tt

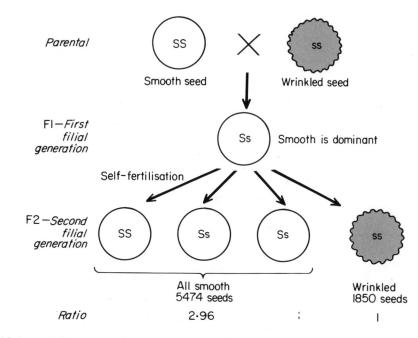

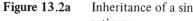

Figure 13.2a   Inheritance of a single character — seed coat texture in the garden pea *Pisum sativum.*

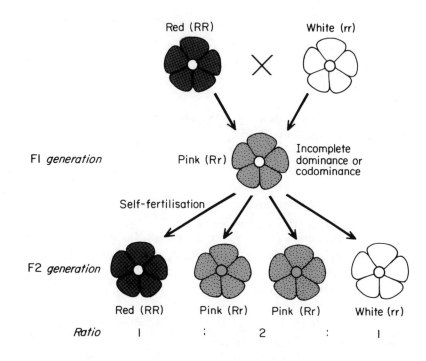

Figure 13.2b   Partial, incomplete or codominance in snapdragon and Marvel of Peru.

(g) Back-cross

This is used to find the identity or gene make-up of an individual, which shows the dominant character in its phenotype but may be homozygous or heterozygous. Back-cross involves crossing the individual to be identified genetically with the *recessive* homozygous parent:

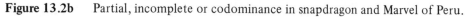

1. *Homozygous hybrid* crossed with a recessive homozygous parent produces 100% phenotypes all heterozygous with dominant character.
2. *Heterozygous hybrid* crossed with a recessive homozygous parent produces 50% phenotypes homozygous with recessive character and 50% heterozygous with dominant character.

(h) Sex Inheritance

In *mammals* females possess an *XX* chromosome pair and males possess an *XY*, or *YX*, chromosome pair.

Female mammals produce 100% *X* ova

Male mammals produce spermatozoa

	X	X
50% X	XX female	XX female
50% Y	YX male	YX male

Therefore offspring can be 50% male, *YX*, or 50% female, *XX*.

The symbol for a male gamete is ♂ and for a female gamete ♀.

(i) Evolution

1. *Variation* in a species arises from changes in genetic composition by *meiosis*; by random pairing in *fertilisation*; and by *mutations* (gene or chromosome changes).
2. *Populations* are groups of same species, and are in homeostasis with their environment.
3. Changes in environment will allow species with certain advantageous variations to survive and reproduce. For example, a sooty environment allowed the dark (melanic) form or variation of the peppered moth to survive predation from birds in cities with air pollution, whereas the white form of moth was easily visible to predators against soot-blackened tree trunks.
4. *Natural selection* is the action of the natural external environment in promoting certain genotypes to survive and reproduce.
5. *Artificial selection* is the action due to human beings promoting certain genotypes of agricultural plants and animals.
6. *Speciation* is the formation of new species through barriers which prevent two populations from interbreeding. They then become progressively different and form *subspecies,* and finally different *species.*
Note. The *characteristics* of a population can change as a result of natural or artificial selection.

(j) Genetic Engineering

Genetic engineering is the moving of DNA sections or genes from the chromosome of one bacterium or virus to another of a different species. This change can result in increased production of certain hormones or vitamins. See page 249.

13.2 Practical Work

(a) Variation

Quantitative measurements should be made to illustrate the following:
(a) *Continuous* variation by measuring in millimetres the lengths of 100 broad bean seeds, and indicating the results on histograms.
(b) *Discontinuous* variation of, for example, eye colour or presence or absence of ear lobes, in a group of people is recorded on a histogram.

(b) Genetics

The *random mating* in gene pair formation is demonstrated by placing 50 black beads (gene *A*, sperm) and 50 white beads (gene *a*, ova) together in one box, and mixing. Two beads are selected randomly and the 'gene pair' combination *AA*, *aa* or *Aa* is recorded in tabular form in three columns. After all beads have been paired, the ratio of phenotypes is determined.

13.3 Examination Work

(a) Multiple-choice Objective Questions

1. The process of meiosis or reduction-division occurs in the cells of:
   (a) plant apical tissue
   (b) mammal ovaries
   (c) plant zygotes
   (d) mammal epithelial cells

2. The fertilised egg has the same number of chromosomes as the parents' body cells; this condition is called:
   (a) haploid
   (b) diploid
   (c) triploid
   (d) polyploid

3. The process of meiosis will normally produce nuclei with a single set of unpaired chromosomes which are called:
   (a) haploid
   (b) diploid
   (c) triploid
   (d) polyploid

4. Genes are known to be located in the nucleus
   (a) nuclear membrane
   (b) nucleolus
   (c) chromosomes
   (d) nuclear fluid

5. Which of the following combinations of *three* substances are normally found in chromosomes of eukaryote cells:
   (i) ATP, (ii), ADP, (iii) protein, (iv) DNA, (v) RNA, (vi) glycogen?
   (a) i, ii, vi
   (b) iii, iv, v
   (c) iv, v, vi
   (d) i, iii, iv

6. Which one of the following is a haploid cell?
   (a) neurone
   (b) spermatozoon
   (c) guard cell
   (d) mesophyll cell

7. A gene is best described as:
   (a) a long protein chain molecule
   (b) a unit of inherited material
   (c) an organelle concerned with respiration
   (d) a gamete or reproductive cell

8. An adult male animal has 46 chromosomes in all its body cells. How many chromosomes will be found in the male gamete cell nuclei?
   (a) 23
   (b) 46
   (c) 69
   (d) 11½

9. How many of the 46 chromosomes found in the adult male animal body cells will be called sex chromosomes?
   (a) none
   (b) 2
   (c) all
   (d) 1

10. Genes affecting the same character or trait are located on the same position in homologous chromosomes and are called:
    (a) autosomes
    (b) alleles
    (c) loci
    (d) chromatids

11. Which of the following is associated with Charles Darwin?
    (a) the principles of heredity
    (b) the origin of species
    (c) the structure of chromosomes
    (d) the theory of acquired characters

12. When a pure-bred red-flowered pea is crossed with a pure-bred white-flowered pea, all the daughter plants show red flowers. This is an example of:
    (a) hybrid vigour
    (b) an acquired character
    (c) a dominant trait
    (d) incomplete dominance

13. The four main blood groups detectable in human beings result from the influence of three genes, A, B and O. Which of the following gene pairs is heterozygous?
    (a) AA

(b) BB

(c) AO

(d) OO

14. Which of the following is an example of discontinuous variation?

    (a) range in height of pea plants from 30 to 50 cm

    (b) offspring include two male cats and three female cats

    (c) pea plants grown in darkness are yellow

    (d) adult human body weight ranges from 50 to 95 kg

Questions 15 to 19 refer to the following account of a genetics experiment.
A plant breeder crossed a red-flowered plant with a white-flowered plant and
obtained all pink flowers ($F1$ generation). On self-pollination, these pink-flowered
plants produced the following offspring ($F2$ generation):

    148 plants with red flowers

    300 plants with pink flowers

    154 plants with white flowers

15. If R represents the gene for red and r represents the gene for white, the geno-
    types of the parents which produced all pink-flowered offspring in the $F1$
    generation must have been;

    (a) $Rr \times Rr$

    (b) $RR \times Rr$

    (c) $RR \times rr$

    (d) $rr \times rr$

16. If one of the $F1$ generation pink-flowered plant was crossed with its white-
    flowered parent, the offspring would be expected to be:

    (a) all red-flowered

    (b) all white-flowered

    (c) 3 pink-flowered : 1 white-flowered

    (d) 1 pink-flowered : 1 white-flowered

17. In the $F2$ generation, the ratio of red- : pink- : white-flowered plants is
    approximately:

    (a) 1 : 1 : 1

    (b) 1 : 2 : 1

    (c) 1 : 3 : 1

    (d) 1 : 4 : 2

18. What percentage of the $F2$ generation plants have pink flowers?

    (a) 25%

    (b) 33%

    (c) 40%

    (d) 50%

19. The occurrence of pink flowers in the $F1$ and $F2$ generations is an example
    of:

    (a) recessiveness

    (b) dominance

    (c) incomplete dominance

    (d) pure breeding

20. Which of the following is the smallest in size compared with the others?

    (a) nucleus

    (b) chromosome

    (c) gene

    (d) nucleolus

(b) Structured Questions

✳ Question 13.1

A recessive sex-linked gene prolongs the blood clotting time, resulting in the condition known as haemophilia. Haemophilia frequently appears in the human population, but occurs very rarely in women. The gene responsible is situated on the X chromosome. Use the family tree shown below to answer the questions.

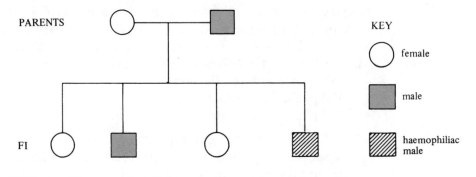

(i) What are the genotypes of the parents?

(ii) Determine the offspring of a marriage between the affected son and a female with no history of haemophilia in her family.

(iii) Determine the offspring of the marriages of the daughters with affected males.

(10 marks)
(NISEC)

✳ Question 13.2

For millions of years, dinosaurs were the dominant animals on Earth. Sixty-five million years ago, however, the age of the dinosaurs ended abruptly and mammals began to dominate. Recent evidence, although not proven, suggests that there was a significant rise in the temperature of the atmosphere at that time. As mammals had better temperature-regulating mechanisms than the dinosaurs, this may have been important.

(a) (i) Give a term which describes the ability of an animal to regulate its body temperature. **(1 mark)**

(ii) What effect would an increase in the external temperature have on the body temperature of a dinosaur? Explain your answer. **(3 marks)**

(iii) Describe the mechanisms by which a named mammal maintains a constant body temperature as the external temperature increases. **(8 marks)**

(iv) Suggest how a sustained increase in the air temperature may have brought about the extinction of dinosaurs, yet enabled the mammals to survive. **(3 marks)**

(b) The mammals that survived the events that caused the extinction of the dinosaurs were able, because of natural selection and their variability, to evolve into the many different kinds of mammal found today.

(i) Explain the importance of meiosis and fertilisation in producing varied offspring.
(5 marks)

(ii) A third cause of variation is mutation. Explain this term. **(1 mark)**

(iii) What is Natural Selection? **(2 marks)**

(iv) If certain species of mammal colonised two very different and isolated habitats, how might natural selection operate to produce two different species? **(5 marks)**

(v) How do fossils help to support the theory of evolution? **(2 marks)**
(AEB, 1983)

(c) **Free-response-type Questions**

(i) *Short-answer Questions*

1. The colour of plumage in Andalusian fowls is determined by a single pair of alleles. One allele (*B*) promotes the production of melanin, and individuals homozygous for this allele are black. Fowls homozygous for the other allele (*b*) are white.

   (a) Using the symbols *B* and *b* to represent the two alleles, complete the table which represents a cross between a black and a white Andalusian fowl.

	(Black parent)	*(White parent)*
Genotype of parent		
Genotype of gamete		
Genotype of *F*1		

   (b) The *F*1 fowls are neither black nor white but blue. What does this suggest about the relationship between the two alleles? **(4 marks)**
   (AEB, 1983)

2. Distinguish between a gene and a chromosome. **(2 marks)**
   (SUJB)

3. In the fruit fly *Drosophila melanogaster* there is a recessive variety in which the wings are reduced to stumps. This variety is called *vestigial-winged*.
   A vestigial-winged fly was mated with a normal-winged one and produced 119 offspring with normal wings and 94 with vestigial wings.

   (a) Using the symbols + for normal wings and *vg* for vestigial wings, give the genotypes of:

   (i) the parents: vestigial-winged parent **(1 mark)**

   normal-winged parent **(1 mark)**

   (ii) the offspring: vestigial-winged offspring **(1 mark)**

   normal-winged offspring **(1 mark)**

   (b) What is the term used to describe this sort of cross? **(1 mark)**
   (OLE)

4. The family tree shows the inheritance of the traits of tongue rolling and inability to roll the tongue.

   A (roller) × B (non-roller)
   |
   C (roller) × D (roller)
   |
   E (non-roller)   F (roller)

   (i) Which is the dominant trait? **(1 mark)**

   (ii) Give the word used to describe the genotype of C. **(1 mark)**

   (iii) For which individual(s) is it *not* possible to state the genotype? **(1 mark)**

   (iv) Is tongue rolling an example of continuous or discontinuous variation? **(1 mark)**
   (SEB)

✻ **Question 13.3**

(a) What are the differences and similarities between mitosis and meiosis? **(6 marks)**

(b) Pollen from a red-flowered plant A was used to pollinate a white-flowered plant B, the seeds from which produced 157 red-flowered plants. When a red-flowered plant C was pollinated by a white-flowered plant D, the resultant seeds gave rise to 84 red-flowered plants and 91 white-flowered plants.

Assume that plants A, B, C and D were all of the same species and that flower colour in that species is controlled by a single pair of genes. Using the symbol R for the dominant gene and r for the recessive gene, write down the genotypes of the following:
  (i) Plant A
  (ii) Plant B
  (iii) The progeny of the A × B cross
  (iv) Plant C
  (v) Plant D
  (vi) The red-flowered progeny of the C × D cross
  (vii) The white-flowered progeny of the C × D cross **(7 marks)**

(c) How is a young mammal before birth (i) protected, and (ii) supplied with raw materials? **(12 marks)**
  (L)

Answer to Q13.3

(a) A *mixed* question. *Similarities* between meiosis and mitosis: both are methods of cell and nucleus division. *Differences*: see the table.

Mitosis	Meiosis
1. One cell divides into *two* daughter cells	1. One cell divides into *four* daughter cells
2. Nucleus chromosome number remains *diploid*	2. Nucleus chromosome number halved, *haploid*
3. Occurs in vegetative growth and repair tissues	3. Occurs in sexual organs with gamete formation
4. No variation occurs in chromosome structure	4. Variation occurs in chromosome structure by crossing over and recombination

(b) Genotypes:
  (i) Plant A, RR – homozygous dominant red
  (ii) Plant B, rr – homozygous recessive white
  (iii) Progeny of A and B cross:

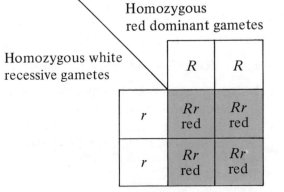

238

All progeny 100% red hybrid heterozygous, *Rr*.
(iv) Plant C, *Rr* − heterozygous red hybrid
(v) Plant D, *rr* − homozygous white recessive
(vi) Progeny of C × D cross:

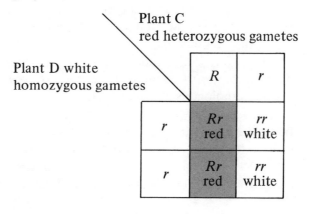

The red-flowered progeny of C × D cross = *Rr* heterozygous red hybrid.
(vii) The white-flowered progeny of C × D cross = *rr* homozygous white.

(c) Figure 12.10 is needed here (Chapter 12).

  (i) *Embryo protection* is by: the mother's *whole* body; the mother's strong muscular *uterus*; the *amniotic* fluid providing *buoyancy* and *hydrostatic* protection from mechanical damage to developing tissue (also acts as a *lubricant* for embryo to move freely, and provides a steady homeostatic external environment).

 (ii) *Raw materials* needed by embryo include nutrients, water, oxygen and protective antibodies. These are contained in the mother's blood, bathing the blood capillaries of the villi in the placenta of the embryo. These raw materials pass by the umbilical vein in the umbilical cord to the embryo's blood circulation.

✽ **Question 13.4**

(a) Explain briefly what you understand by each of the following terms: (i) genotype; (ii) phenotype; (iii) gamete; (iv) back-cross; (v) recessive; (vi) heterozygous. **(12 marks)**

(b) In a genetic experiment a wild rabbit was mated with a tame white rabbit and it was found that all the offspring were brown, like the wild parent. These offspring were then mated with other white rabbits to produce a total of 216 offspring, 103 of which were white and 113 were brown. Explain fully the genetics of these crosses. **(13 marks)**
(OLE)

Answer to Q13.4

(a) (i) *Genotype* is the gene composition of an organism.
 (ii) *Phenotype* is the physical characteristics which can be seen or measured and is the result of the interaction between the gene composition (genotype) and the environment.
(iii) *Gametes* are the reproductive cells formed by meiosis and have *half* the normal number of chromosomes (haploid).

(iv) *Back-cross*, or test-cross is used to *identify a genotype*, i.e. whether it is homozygous or heterozygous, by crossing with the homozygous recessive parent. The homozygous hybrid will produce 100% heterozygous offspring, and the heterozygous hybrid will produce 50% homozygous and 50% heterozygous offspring.

(v) *Recessive* is that member of the pair of genes which does not show itself in the presence of its contrasting gene. It is seen in the phenotype when the two recessive genes or *double recessives* are present together.

(vi) *Heterozygous* means to have two *different* or contrasting genes or alleles on homologous chromosomes.

(b) The wild brown rabbit is assumed to be a pure-line homozygous hybrid, *BB*, whereas the tame white rabbit is a pure-line homozygous hybrid, *bb*. The dominant character, *brown* (*B*), is seen in the *F*1 generation as a consequence of the brown wild-rabbit and white tame-rabbit cross.

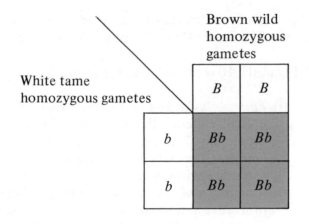

The *F*1 progeny are *all* brown heterozygotes, *Bb*.

The brown heterozygotes, *Bb*, from the *F*1 generation, when mated with other white homozygous rabbits, *bb*, produce the following progeny:

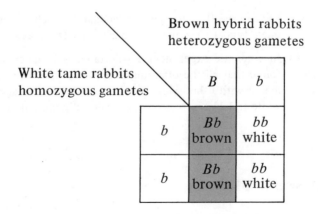

The *F*2 generation consists of 50% brown *Bb* heterozygotes and 50% white *bb* homozygotes, almost in the same ratio (113 brown to 103 white) as stated in the question, i.e. a ratio of 1 : 1.

13.4 Self-test Answers to Objective and Structured Questions

Answers to Multiple-choice Objective Questions

1. b 2. b 3. a 4. c 5. b 6. b 7. b 8. a 9. b 10. b 11. b 12. c
13. c 14. b 15. c 16. d 17. b 18. d 19. c 20. c

Answer to Structured Question 13.1

The haemophilia gene can be shown as *Xh*. Since the condition rarely occurs in the female sex as *XhXh*, the women will have the genotype *XhX* as female *carriers*, and the haemophiliac *affected* male will be *XhY* and normal male will be *XY*.

(i) Parents' genotype will be: female = *XhX* female carrier; male = *XY*

(ii) The *F*1 generation will be: *XhX* or female *carrier*; *XX* or *normal* female; *XY* or *normal* male; *XhY* or *affected* male. The offspring of *affected son XhY* and a *normal* female *XX* is determined as follows:

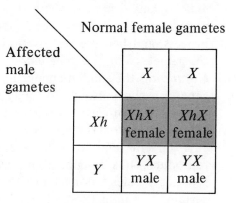

The offspring will be *normal XY* males but *all* the females will be *carriers*, *XhX*.

(iii) The offspring of the marriage of the daughters will be as follows:
   Normal daughter *XX* and *affected* male *XhY* will be the same as shown above, namely all daughters are *carriers XhX* and all sons *normal XY*.
   The marriage of the *carrier* daughter *XhX* and *affected* male *XhY* will produce offspring as follows:

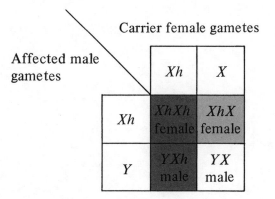

The offspring will be as follows: some daughters are *carriers, XhX*; *half* of the sons are *affected, XhY*; the other sons are *normal, XY*. *Rarely*, some daughters could be affected (*XhXh* haemophiliacs).

Answer to Structured Question 13.2

This is a *mixed* question concerned with the effect of a changing external environment and the extinction of dinosaurs.

(a) (i) Homeostasis, in endotherms or homoiotherms.

    (ii) The dinosaur, being a *reptile* and an *ectotherm* or poikilotherm, would have a body temperature a few degrees above that of the external temperature.

    (iii) Mammals are *endotherms* or homoiothermic, and maintain a constant body temperature as the external temperature increases by: *vasodilation* of *skin* capillaries and heat radiation; *sweat* secretion and evaporation, drawing latent heat from the body; *hair* erector muscles relax and hair lies flat, allowing heat radiation; *panting* to encourage water evaporation by lungs and tongue. There are also *behavioural* means, for example withdrawing to shade or bathing. *Dogs* use these mechanisms.

    (iv) A sustained increase in air temperature caused increased metabolism rates in reptiles, with increased enzyme activity until the higher temperature destroyed the enzymes at temperatures above the optimum.

(b) (i) *Meiosis* and *fertilisation* are the cause of variation in offspring, because meiosis produces a new combination of genes through the crossing-over process; and because fertilisation produces *random* mating of genes from the gametes and new *gene pairs*.

    (ii) *Mutation* is the sudden change in the composition or structure of either a gene or a chromosome and may occur naturally in meiosis or artificially by the effect of certain drugs or radiation.

    (iii) *Natural selection* is a process by which organisms less adapted to their environment tend to become extinct and the better adapted survive to reproduce, thus bringing about *evolution*.

    (iv) Existing species colonising different and isolated habitats could develop certain characteristics adapted to these environments as a result of natural selection; this would result in new species or subspecies *unable* to reproduce with the other species. This is called *speciation*.

    (v) *Fossils* are remains or impressions of long-dead plants and animals, providing evidence of prehistoric forms of life from which present life has evolved.

Theme IV

Relationships between Organisms and with the Environment

14 The Ecosystem

14.1 Theoretical Work Summary

(a) Ecology

Ecology is the study of living organisms in relation to their environment, the surroundings in which organisms live.

The external environment has living (*biotic*) and non-living (*abiotic*) components. The former comprise the plants and animals living in the Earth's *biosphere* (sea, land and air). The latter constitute the physical and chemical factors in the environment of temperature, water, light, minerals, pH and soil.

The internal environment is the *tissue fluid* surrounding tissues and cells (see Chapter 9, Homeostasis, Section 6.3g, Fig. 6.4 and Section 9.1a).

A natural habitat or ecosystem, either a *deciduous tree* or a *freshwater pond*, should be studied with regard to the following headings (b) to (g).

(b) Ecosystem Composition

The *ecosystem* comprises the non-living environment, e.g. lakes, oceans and forests, and the organisms that live in it. Members of one species are termed *individuals*, and the total number of such members of the same species in a given area at the same time is described as a *population*. The different interacting populations of species in the same environment or habitat are a *community*. Ecosystems are thus the *interactions* between the community and the non-living environment.

(c) Major Living Components of Ecosystems

The three main biotic components of ecosystems are producers, consumers and decomposers.

1. *Producers* are *autotrophic green* plants (trees, shrubs and herbs) (see Chapter 4).

$$\text{SUNLIGHT ENERGY} + 6CO_2 + 6H_2O \longrightarrow \underset{C_6H_{12}O_6}{\underset{\text{(glucose)}}{\text{Carbohydrate}}} + \underset{6O_2}{\text{Oxygen}}$$

2. *Consumers* (heterotrophs) obtain energy by feeding on producers as primary consumers (*herbivores*), or as secondary consumers as *omnivores* or *carnivores*.

3. *Decomposers* (mainly fungi and bacteria) feed on *dead* organic matter to obtain energy. *Fungi* feed mainly on *cellulose* (wood and cell walls), and *bacteria* mainly break down *protein*. (See Section 5.1.)

(d) Food Chains and Webs

The *food chain* is a feeding relationship in which energy in carbon compounds from producers is transferred to other consumers at different trophic levels. *Food webs* (Fig. 14.1) are composed of interconnected food chains. *Pyramids of biomass* show the amount of total dry weight of living material at each trophic level of a food chain. *Pyramids of numbers* show the numbers of living organisms at each trophic level.

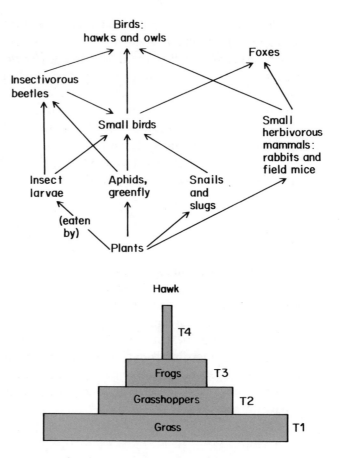

Figure 14.1 A food web composed of interrelated food chains.

(e) Ecosystem Energy

Energy *flows* through or is transferred through the trophic levels and living components of ecosystems; over 90% is lost as heat at each level to pass on to outer space. (See Fig. 14.2.)

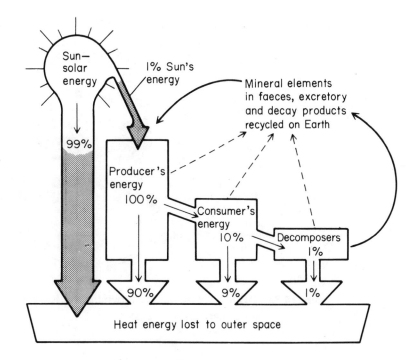

Figure 14.2   Energy flow through the biosphere.

Fossil fuels, namely coal, oil and gas, which form from organic plant and animal remains, are valuable *energy sources*, releasing their energy on *combustion*.

(f) Ecosystem Recycling of Chemical Elements

Chemical elements are changed into chemical compounds in living organisms using energy. The same chemical elements are *naturally recycled* for the continuation of life on Earth.

1. *Carbon* recycled by: respiration, decay, combustion, photosynthesis and fossilisation.
2. *Nitrogen* recycled by: nitrification, denitrification, fixation by root-nodule *bacteria*, chemical synthesis (lightning), leaching and decay.
3. *Water* recycled by: transpiration, evaporation, respiration, perspiration, combustion and condensation.
4. Human beings are agents in recycling glass, certain metals and paper.

(g) Interrelations between Organisms

In addition to the main 'producer–consumer–decomposer' relationship, the following also exist.

1. *Symbiosis* is a beneficial partnership between different bacteria in their root nodules, lichens (a partnership between an alga and a fungus), and bacteria in the human gut forming vitamins of the B group.
2. *Parasitism* is a relationship between *parasite* and *host* in which the parasite *benefits* and the host is *harmed*. Examples of *ectoparasites* are dodder on gorse (Fig. 14.3a); ringworm; fleas and lice on skin; and greenfly on plants. Roundworms and tapeworms (see Fig. 14.3b) are *endoparasites*.
3. *Competition* is a relationship between organisms of different species in which they compete for *resources* (food, water, raw materials). *Human beings* are the most savage competition in the biosphere, affecting the environment by:

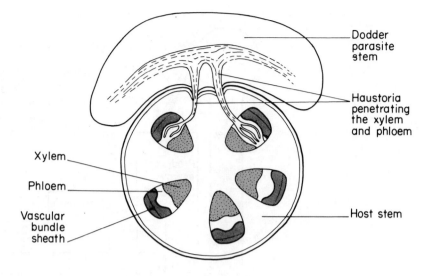

Figure 14.3a   Dodder, *Cuscuta* sp., a twining rootless ectoparasite on nettles, gorse and heather.

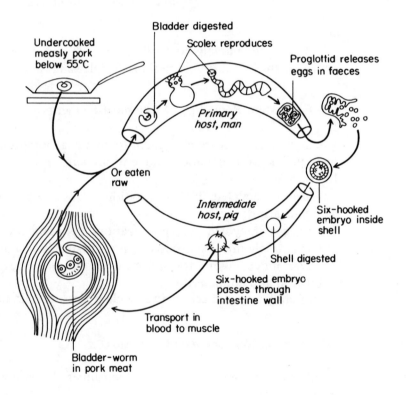

Figure 14.3b   Life cycle of the pork tapeworm, *Taenia solium*, an endoparasite of man and the pig.

    (i) *elimination* of species, by over-hunting, fishing and use of *pesticides* and *herbicides*;
    (ii) *habitat* destruction in land clearance;
    (iii) *pollution* of water, air and land by *industrial* activities, *radioactive* contamination, and consumption of *irreplaceable* fossil fuels and mineral resources.

(h) Human Beings and Disease

The fact that the world's population has now reached over 4000 million is due partly to the conquest of the diseases caused by *bacteria*, *fungi*, *protozoa* and *viruses* (microbial diseases; see Table 14.1) and partly to improved nutrition through increased food production, leading to a decrease in the *infant death rate*.

Mammalian Body Defences

1. *Immunity* is conferred by means of protein *antibodies* made by the blood *lymphocytes* which react with *antigens* made by pathogenic microbes.
2. Skin, mucous membranes, blood, lymph and the liver are defences summarised in Fig. 14.4.

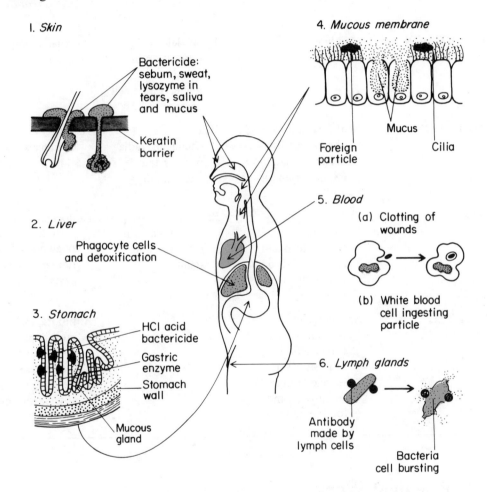

Figure 14.4 Natural body defences against infectious disease.

(i) Biotechnology

Biotechnology is the *large-scale* production by certain bacteria or fungal *micro-organisms* of substances beneficial to human beings. Genetic engineering (page 232) can be involved so as to produce maximum amounts of these substances which are by-products of micro-organism metabolism. A special form of *insulin* (page 163), *growth hormone* (page 178), natural *insecticides*, and bacterial *protein* are some of the products of biotechnology.

Table 14.1   Summary of microorganisms

	Bacteria	Viruses	Fungi (moulds)	Protozoa
STRUCTURE	Unicellular; DNA strand; no definite nucleus; 100–2000 μm	Non-cellular; particles or virions; DNA or RNA strand; 10–500 μm	Unicellular thread-like *hyphae*; definite nucleus;	Unicellular with definite nucleus; over 500 μm
NUTRITION	Mainly saprophytes, parasites and a few symbionts. Feed mainly on *protein*	Parasites on *living* cells, which produce the protein needs of the viruses	No chlorophyll; mainly saprophytes or parasites feeding on *carbohydrate*	Free-living in water or are parasites feeding on various organisms phagocytically
REPRODUCTION	Mainly asexual	By nucleic acid replication in living host nuclei. Cause cell damage to host	Asexual mainly by *spores*	Asexual mainly, and some sexual
DESTRUCTION	Heat (dry and moist); chemicals (disinfectants); radiation (X and gamma); antibiotics	Heat; chemicals; radiation; certain anti-viral drugs	Heat (dry); chemicals (fungicides); antibiotics	Various antiprotozoal drugs
HARMFUL	*Pathogens*, causing such diseases as tuberculosis and cholera, and food poisoning; denitrifiers	*Pathogens*, causing such diseases as influenza and smallpox	*Pathogens*, causing such diseases as ringworm. Food, timber, textiles — pests	*Pathogens*, causing such diseases as amoebic dysentery, malaria
BENEFICIAL	Antibiotic and vitamin B group formers. Soil nitrogen fixers, sewage protein decomposers, and food fermenters	None	Antibiotic and protein producers. Soil humus formers, sewage cellulose decomposers, and food fermenters	Sewage detritus feeders

14.2 Practical Work

(a) Ecology

(i) *Methods for Studying a Field or Hedgerow Habitat*

These include the use of:
1. *Quadrat*, a square frame of metal or wood, 1 m x 1 m or 0.5 m x 0.5 m, to determine distribution of various species in randomly selected parts of an area.
2. *Transect line* made by a long piece of line between two poles. A record is made of every species touching the line in a line transect. A belt transect includes all species within 0.5 or 1 m of the line.

(ii) *Collection Methods*

Those of importance are the pitfall trap and the Tullgren funnel for small arthropods in soil samples (Fig. 14.5). *Tree beating* should be demonstrated by collecting organisms on sheets or inverted umbrellas.

Pond specimens are collected by means of small- and large-mesh nets.

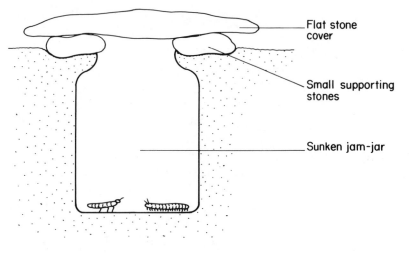

Figure 14.5a   Pitfall trap for collecting small soil animals.

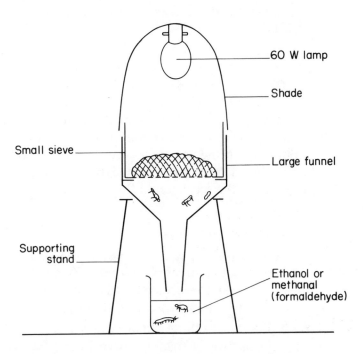

Figure 14.5b   Tullgren funnel for collecting small animals from soil.

Note. Collection of organisms must be done with *care*, so that there is no damage and little or no disturbance to the specimens or their habitat.

(b) Soil

The study of soil components is related mainly to experimental work, as follows.
1. Soil Moisture, Mineral (Inorganic) and Organic Components:
   (a) *Moisture*: dry a *known weight* of fresh soil sample to constant weight at 110°C; determine weight loss.

(b) *Organic content*: heat all the sample of moisture-free soil strongly in an evaporating basin to a constant weight over a bunsen burner. Record weight loss after *strongly* heating the dry soil.

(c) *Inorganic mineral content* is the final weight of soil remaining after strong heating.

$$\text{Water content, \%} = \frac{\text{Weight loss after heating at } 110^{\circ}\text{C}}{\text{Weight fresh soil}} \times 100$$

$$\text{Organic humus content, \%} = \frac{\text{Weight loss after burning DRY soil}}{\text{Weight fresh soil}} \times 100$$

$$\text{Mineral inorganic content, \%} = \frac{\text{Weight of soil residue after reducing to ash}}{\text{Weight fresh soil}} \times 100$$

(All percentages are based on *fresh soil* composition.)

2. *Physical analysis* of mineral components of a soil is carried out by stirring soil sample in water in a tall glass jar or cylinder (Fig. 14.6).

3. *pH measurement* involves mixing a small sample of soil with a little distilled water and adding universal indicator, allowing mixture to settle and comparing the water colour with colours shown on the pH comparator card.

4. *Micro-organisms* can be demonstrated by inoculation of sterile culture media with fresh soil. The control is inoculated with *sterile* soil produced after strong heating (see Fig. 14.7).

(c) Organism Examination

Living organisms obtained by different collecting methods should be carefully examined and their *external features* drawn; afterwards the living specimens should be returned to their place in the habitat. The following specimens should be examined from each named habitat (the page numbers and figures refer to diagrams of the specimen in this text).

Freshwater Pond Habitat

Herbaceous plant: any flowering aquatic plant.
Alga: filamentous green *Spirogyra* or *Chlamydomonas*; see page 25.
Insect: any water insect such as a water beetle.
Pond snail for molluscan features; see page 31.
Bony fish such as the stickleback; see page 34.
Amphibian: frog or toad; see page 34 and Fig. 8.10.
Protozoa; see page 29.
Bacteria from prepared microscope slide; see page 24.

Deciduous Tree

Tree twig in winter, in summer, and fruit. See Figs 4.1, 8.7, 12.3 and 12.7.
Herbaceous flowering plant; see page 27.
Alga found on tree trunk, e.g. *Pleurococcus*: examine a prepared microscope slide.
Fungi: examine microscope slide of *Penicillium* (see page 27), or a field mushroom.
Bacteria; see page 27.
Earthworm; see page 31.

Figure 14.6   Separation of solid components of soil by physical analysis.

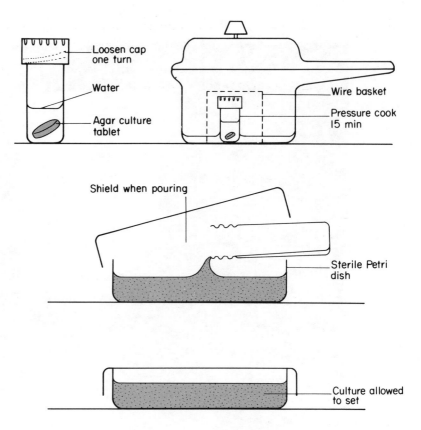

Figure 14.7 Apparatus used in preparing a culture media plate.

Snail: see page 31.
Insect, e.g. butterfly or moth; see Fig. 8.9.
Bird: wood pigeon; see page 35.
Mammal: squirrel or rabbit; see page 35.

14.3 Examination Work

(a) Multiple-choice Objective Questions

1. Compared with a clay soil, a sandy soil:
   (a) has good water-retaining property
   (b) is light and easy to dig
   (c) is slow draining
   (d) has small, limited air spaces

2. Higher plants do not benefit by the activities of the nitrogen-circulating bacteria in the soil that cause:
   (a) breakdown of nitrates into nitrogen
   (b) nitrogen fixation from the air
   (c) oxidation of nitrites to nitrates
   (d) conversion of amino acids into ammonia

3. Which of the following food chains is the most efficient for making energy available to human beings, assuming the same mass of green plant material is consumed in each case?
   (a) green plant ⟶ sheep ⟶ human beings
   (b) green plant ⟶ human beings

(c) green plant $\longrightarrow$ zooplankton $\longrightarrow$ fish $\longrightarrow$ human beings

(d) green plant $\longrightarrow$ birds $\longrightarrow$ eggs $\longrightarrow$ human beings

4. The process of maintaining cells, tissues or bacterial cultures at a constant temperature of about 38°C is called:
   (a) inoculation
   (b) incubation
   (c) sterilisation
   (d) fertilisation

5. The organic matter found in topsoil is called:
   (a) humus
   (b) silt
   (c) loam
   (d) sand

6. The method of feeding of the pin mould *Mucor* on stale bread or fruit is described as:
   (a) holophytic
   (b) saprophytic
   (c) parasitic
   (d) autotrophic

7. State which of the following is the *producer* in a food chain composed of:
   (a) ladybirds
   (b) aphids (greenfly)
   (c) rose bush
   (d) blue tit

8. Which of the following soil micro-organisms changes protein in plant and animal remains into nitrates?
   (a) denitrifying bacteria
   (b) nitrifying bacteria
   (c) fungal moulds
   (d) nitrogen-fixing bacteria

9. A fully equipped and self-contained cargo ship carrying a range of animals from Africa to stock a zoo in Europe is an example of:
   (a) a population
   (b) a community
   (c) a niche
   (d) a food web

10. The dodder plant *Cuscuta* lives on the surface of gorse, *Ulex* sp., by penetrating the bark and drawing nutrients from the vascular tissue. The relationship between the dodder plant and gorse is called:
    (a) mutualism
    (b) predation
    (c) parasitism
    (d) competition

11. All members of the same species found in a particular community are called:
    (a) a niche
    (b) an ecosystem
    (c) a population
    (d) a family

12. In the food web of the sea the producers obtain energy from the:
    (a) zooplankton
    (b) sun
    (c) sea water
    (d) air

13. The correct order in which soil particles separate after stirring in water is shown as follows, with the *larger* particles named first.
    (a) sand, gravel, clay and silt
    (b) gravel, sand, silt and clay
    (c) silt, clay, gravel, sand
    (d) clay, sand, silt, gravel

14. The water-retaining properties of a soil are improved by adding:
    (a) peat
    (b) lime
    (c) sand
    (d) inorganic fertiliser

15. Which of the following air pollutants is a factor in causing bronchitis:
    (a) carbon dioxide
    (b) carbon monoxide
    (c) sulphur dioxide
    (d) nitrogen oxides

16. German measles, smallpox and influenza are all diseases caused by:
    (a) fungi
    (b) protozoa
    (c) viruses
    (d) bacteria

17. Which of the following is a method of food preservation involving sterilisation?
    (a) refrigeration
    (b) smoking
    (c) canning
    (d) sun-drying

18. The microscopic living threads which make up the body of a mould fungus are called:
    (a) hyphae
    (b) cilia
    (c) flagella
    (d) spores

19. Penicillin is an antibiotic used to inhibit the growth of certain micro-organisms causing conditions such as:
    (a) staphylococcal sore throat
    (b) viral influenza
    (c) fungal ringworm
    (d) protozoan dysentery

20. The apparatus or method used to collect microscopic arthropods from a soil sample is called:
    (a) a pitfall trap
    (b) the Tullgren funnel apparatus
    (c) incubation on nutrient culture media
    (d) soil air analysis

* (b) Structured Questions

Question 14.1

Two species of mite were kept in the laboratory. One species, Mite A, feeds on oranges. The other, Mite B, feeds on Mite A. The table shows how the populations of the two mites varied with time.

Time (days)	Numbers of Mite A	Numbers of Mite B
5	50	—
10	250	—
15	400	—
18	500	50
22	1400	200
28	1900	450
30	950	1050
35	850	1200
40	600	1600
42	0	600
47	0	50
52	0	50
55	0	0

(a) Plot graphs of the information given in the table. **(7 marks)**

(b) Comment on the situation revealed by the graphs with regard to the levels of predator and prey populations. **(4 marks)**

(c) When the predators are too efficient, what happens to both populations? **(1 mark)**

(d) If predators were entirely removed, what might happen to the prey population? **(3 marks)**
(SUJB)

* Question 14.2

The main source of protein for cattle is the grass they eat. To ensure a constant supply of grass and thus protein, farmers apply fertilisers to the soil. The table shows the effect on protein yield of a grass of different rates of application of a nitrogen fertiliser.

Total nitrogen applied during season (kg/ha)	Protein yield of grass (as % of dry matter)
0	14.2
112	15.0
224	16.2
336	17.8
448	20.6
672	22.1
896	23.4

(a) (i) Plot a graph of these results. Show rate of application on the horizontal (x) axis and protein yield on the vertical (y) axis. **(6 marks)**

(ii) Which increase in rate of application of fertiliser produced the greatest percentage increase in protein yield? **(1 mark)**

(iii) How much fertiliser must a farmer apply to achieve a 19.2% yield of protein? **(2 marks)**

(b) The table simply refers to 'total nitrogen'.
(i) Name *two ions* containing nitrogen which may be present in a nitrogen fertiliser. **(2 marks)**

(ii) How are these ions absorbed by the grass? **(2 marks)**

(iii) State *four* ways in which the nitrogen absorbed by the grass could be returned to the soil. **(4 marks)**

(iv) Describe *one* other way by which ions containing nitrogen could enter the soil. **(2 marks)**

(v) How could the nitrogen be made available to the grass? **(2 marks)**

(c) Unlike animals, which eat 'ready-made' protein, green plants manufacture their own.
(i) Name the *four* elements present in all proteins. **(1 mark)**

(ii) Name the chief synthetic process occurring in green plants. **(1 mark)**

(iii) The process you have named does not at first manufacture proteins. What class of food does it manufacture? **(1 mark)**

(iv) Describe how green plants manufacture proteins. **(6 marks)**
(AEB, 1982)

* Question 14.3

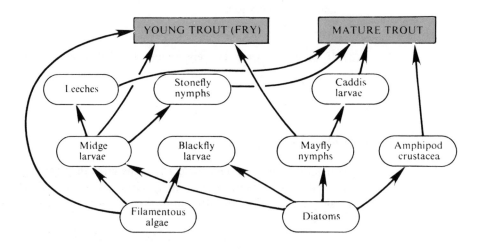

(a) The diagram shows a simplified food web involving trout at different stages of its life history.
(i) The ultimate source of energy for all the organisms in the food web is the Sun. Describe *one* route by which the Sun's energy may be made available for movement in trout. In your answer you should make reference to any processes important in the transfer of energy. **(8 marks)**

(ii) Using examples from the above food web to illustrate your answer, explain what is meant by each of the following terms: A, producer; B, primary consumer; C, secondary consumer. **(6 marks)**

(iii) Describe *two* differences in the diets of young and mature trout. **(2 marks)**

(iv) Suggest a possible effect on mature trout of a disease killing all mayfly nymphs. Explain your answer. **(3 marks)**

(b) The rivers in which fish such as trout live may become 'heat polluted' by power stations discharging warm water into them. The table shows the relationship between water temperature and its ability to dissolve oxygen.

Water temperature (°C)	Amount of oxygen dissolved (mg/litre)
0	14.5
5	12.5
10	11.0
15	10.0
20	9.0
25	8.0
30	7.5

(i) Trout are said to be 'cold blooded'. Explain what is meant by this term. **(2 marks)**

(ii) What effect would an increase in water temperature have on the metabolic rate of trout? **(1 mark)**

(iii) Describe the relationship between water temperature and its ability to dissolve oxygen. **(1 mark)**

(iv) What effect would an increase in water temperature have on the trout's demand for oxygen? Explain your answer. **(3 marks)**

(v) What effect will an increase in water temperature have on the availability of oxygen? **(1 mark)**

(iv) Explain what may happen to trout as a result of your answers to (iv) and (v). **(3 marks)**

(AEB, 1982)

(c) Free-response-type Questions

(i) *Short-answer Questions*

1. Describe how you would find the percentage of humus in a sample of fresh soil. **(5 marks)**

(SEB)

2. (i) Explain concisely one method of food sterilisation. Why is this method of preservation effective?
   (ii) What other methods may be used to preserve food? (Give the biological reasons for each method.) **(6 marks)**

(NISEC)

3. (a) Clotting of blood and phagocytosis are two of the defence mechanisms found in mammals. Explain how each helps to prevent a disease-causing organism from becoming established in a mammal's body.
   (b) What further defence mechanism exists should the disease-causing organism become established? **(4 marks)**

(AEB, 1983)

4. Give a short explanation for the following observation. The total population of plants in any community of organisms is usually far greater than that of the animals. Among the animals there are more herbivores than carnivores.

(5 marks)
(SUJB)

(ii) *Long-answer Questions*

* **Question 14.4**

(a) How are foreign organisms prevented from entering the body of a mammal? **(4 marks)**
(b) If disease-causing organisms succeeded in evading these defences, what responses are made by the tissues? **(6 marks)**

(c) How can the body's natural defences be aided by medical science? **(5 marks)**
(SUJB)

Answer to Q14.4

(a) The *outer barrier* of a mammal includes the *skin* and its tough impermeable layer; and *mucus* from lung and nose epithelia aided by cilia in trachea epithelium, which traps small particles of dust and bacteria.
(b) *Tissues* respond in body defence through the following:
    (i) *blood clotting* to prevent blood loss and further entry via open wounds; *blood phagocytes*, which engulf and destroy bacteria in blood;
    (ii) *lymph* cells called lymphocytes produce *antibodies* acting against antigens of micro-organisms, and provide *natural immunity*;
    (iii) *secretions* of sweat, gastric juice and tears contain *bactericidal* substances destroying bacteria;
    (iv) *liver* has phagocytes and also detoxifies poisonous substances.

(c) *Medical science* aids the body's natural defences by:
    (i) *Vaccination* against certain diseases such as diphtheria, tetanus, whooping cough, TB and polio, using vaccines which provide artificially acquired *immunity*.
    (ii) *Antiseptics* are used on the skin and destroy *most* harmful micro-organisms.
    (iii) *Antibiotics* produced by fungal moulds and bacteria, penicillin and tetracyclines treat many fungal and bacterial infections.
    (iv) *Antiviral* preparations include *interferon*, which prevents virus growth.
    (v) *Antifungal* preparations act against ringworm.
    (vi) *Other* preparations include drugs acting against parasitic worms and protozoa.

* **Question 14.5**

(a) Make a list of the components of a fertile soil and discuss the importance of each.
(9 marks)

(b) What is the significance of: (i) practising crop rotation; (ii) adding lime to soil. **(6 marks)**
(SUJB)

Answer to Q14.5

(a)

Soil component	Importance
MINERAL COMPONENTS	Size of particle determines the soil character, texture and structure — sandy, clay, loam, etc.
SOIL WATER	Contains dissolved substances such as mineral ions and gases which are absorbed by the plant root hairs. Water is an essential requirement for plant structure and transpiration
SOIL AIR	Present between soil mineral particles, essential for root-hair *aerobic respiration*, and for soil micro-organisms
SOIL pH	*Acid* soils are poorly drained and contain peat. *Alkaline* soils are associated with chalk and lime-stone. Plants are unable to tolerate extreme ranges of pH
HUMUS	The main organic component of soil with important water-holding properties for sandy soils, and source of organic nitrogen; important in aerating and draining *clay* soils
MICRO-ORGANISMS	Include the important nitrogen fixers, and nitrifying bacteria, together with fungal moulds and bacteria that form humus as a product of plant and animal decay
OTHER ORGANISMS	Earthworms and moles burrow and aerate soil; ingest, grind and neutralise soil acidity; wormcasts provide good tilth. Soil is an important habitat for many different organisms

(b) *Crop rotation* is the rearing of a different crop on the same land area. Shallow-rooting cereal crops such as wheat are followed by deeper-rooting crops (potatoes) in the second year, and *legumes* (i.e. peas, beans or clover), which have nitrogen-fixing bacteria in root nodules, in the third year. *Cattle* may graze on clover and provide a surface distribution of manure as a mulch.

(c) *Liming*, the addition of calcium oxide or calcium hydroxide or calcium carbonate to soil, improves *soil structure*:
  (i) Heavy clay soil particles form crumbs by *flocculation* to improve aeration and drainage.
  (ii) *Soil acidity* is neutralised, encouraging growth of nitrogen-fixing and nitrifying bacteria.
  (iii) *Mineral ion release*: certain valuable minerals, e.g. potassium and magnesium, are released from the mineral components of soil in exchange for the calcium ions.

✳ **Question 14.6**

  (a) Name the major living components of ecosystems. **(4 marks)**

  (b) By means of a diagram illustrate the flow of energy through an ecosystem. **(3 marks)**

(c) Write a chemical equation for the process by which energy enters an ecosystem.

(6 marks)

(d) Name and describe the effects on the environment of each of: (i) three air pollutants; (ii) three water pollutants.

(12 marks)

(L)

Answer to Q14.6

(a) Since an *ecosystem* comprises a community of living organisms and the non-living environment, the *living components* will include:

  (i) *producers* or green plants which synthesise organic substances by *photosynthesis* (other organisms may also be included, which are bacteria able to form organic materials by *chemosynthesis*);

  (ii) *consumers*, mainly animals and certain parasitic and insectivorous *plants*, which feed on green plants as primary consumers (the carnivores are *secondary* consumers);

  (iii) *decomposers*, which are organisms (bacteria, fungi, earthworms and certain insects) feeding on dead organic matter.

(b) Figure 14.2 can be shown here in a simpler form.

(c) Energy enters the living ecosystem to be trapped as sunlight energy in photosynthesis.

$$\text{SUNLIGHT ENERGY} + \text{carbon dioxide} + \text{water} \xrightarrow{\text{CHLOROPHYLL}} \text{glucose} + \text{oxygen}$$

The chemical equation is:

$$\text{SUNLIGHT ENERGY} + 6CO_2 + 6H_2O \longrightarrow C_6H_{12}O_6 + 6O_2$$

(d) (i) Three air pollutants include: *lead* from petrol, which accumulates in blood of humans, and may affect mental development of children; *sulphur dioxide* from coal burning is harmful to lung tissue of human beings, causing the lung disease bronchitis and related heart disorders, and it may harm the leaves of certain plants; *carbon dioxide* from combustion of fossil fuels may be causing an increase in world temperature due to the 'greenhouse effect'.

  (ii) Three water pollutants: *heat* from power station and factory cooling waters reduces oxygen availability and destroys fish at temperatures above $30^\circ C$; marine seawater pollution by *oil* spillage injures sea birds; *mercury* is a volatile liquid element which evaporates into the air, and also gathers in rain water and collects in factory effluents: increasing amounts of mercury are recorded in freshwater and marine fish (this metal is *accumulated* in tissues and not excreted).

14.4 Self-test Answers to Objective and Structured Questions

Answers to Multiple-choice Objective Questions

1. b   2. a   3. b   4. b   5. a   6. b   7. c   8. b   9. b   10. c   11. c   12. b
13. b   14. a   15. c   16. c   17. c   18. a   19. a   20. b

Answer to Structured Question 14.1

(a) The graph is shown in the accompanying figure.

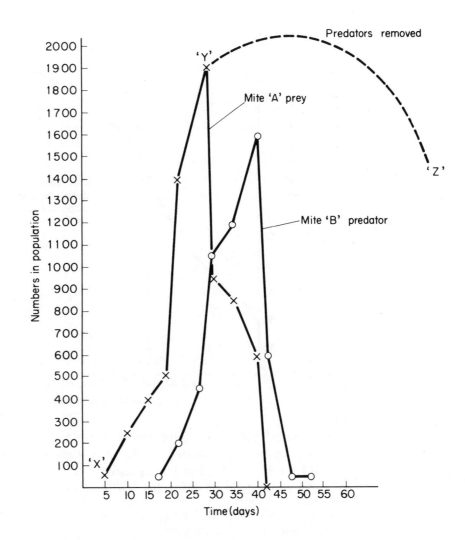

(b) As the levels or numbers or the prey Mite A *increase*, the predators begin to multiply and also increase.

(c) When predators are too efficient, they destroy more prey and their own numbers drop because of lack of prey on which to feed.

(d) The prey population would increase along the curve 'X–Y', when the predators were entirely removed, following maximum rate of population increase due to abundant food supply. When the food supply is exhausted, and excretory waste accumulates, the population will follow the projected curve Y–Z due to death of the population as a result of starvation and pollution.

Answer to Structured Question 14.2

(a) (i) Graph is shown in the diagram which follows.

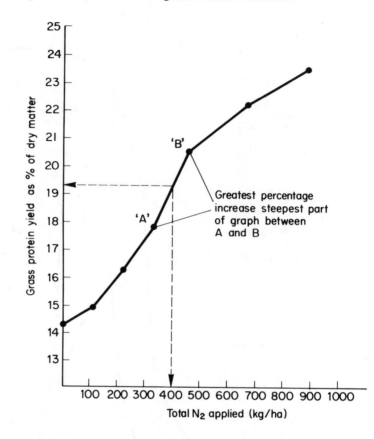

(ii) Greatest percentage increase in protein yield occurred between 336 to 448 kg/ha when protein yield rose from 17.8% to 20.6%, or by (2.8 ÷ 17.8 × 100 = 15.73%. This is also the steepest part of the graph.

(iii) An application of about 400 kg/ha will give a 19.2% yield of protein.

(b) (i) Ions could be NO_3^- (nitrate) or NH_4^+ (ammonium).

(ii) Absorption by root hairs.

(iii) Decomposition of grass to humus; urine of primary consumer (sheep); decomposition of body of primary consumer (sheep); urine of secondary consumer animal; man eats sheep; or decay of human body.

(iv) By *nitrates* formed by chemical synthesis (lightning) or chemical fertilisers.

(v) *Nitrification* is the process of converting complex organic nitrogen-containing compounds into *nitrates* easily absorbed by the grass.

(c) (i) Carbon, hydrogen, oxygen and nitrogen.

(ii) Photosynthesis.

(iii) Carbohydrates.

(iv) *Protein synthesis* is the process of protein manufacture, which occurs in the *ribosomes* of the cytoplasm, using *amino acids* formed by combination of simple carbohydrate with ammonia from absorbed nitrates. The DNA in the nucleus gives instructions to RNA for combining different amino acids to make the different structural and functional proteins:

Nitrate and ammonia + carbohydrate molecules ⟶ amino acids

Different amino acids — Condensation in the ribosomes → different plant proteins

Answer to Structured Question 14.3

(a) (i) The Sun's energy is made available for *movement* in trout by *photosynthesis* in filamentous algae and in diatoms, both of which contain chlorophyll and form carbohydrate from carbon dioxide, water and sunlight energy.

$$\begin{array}{ccc}
\text{Carbon dioxide} + \text{water} & \xrightarrow[\substack{\text{SUNLIGHT}\\\text{ENERGY}}]{\substack{\text{CHLOROPHYLL}\\\text{AND}}} & \text{glucose} + \text{oxygen} \\
6CO_2 \quad\quad + 6H_2O & & C_6H_{12}O_6 + 6O_2
\end{array}$$

The glucose is a compound with a *high energy value*; it reaches the trout by the food chain. The energy is released as ATP (adenosine triphosphate) by aerobic respiration; this causes muscle contraction and therefore movement in the trout.

$$C_6H_{12}O_6 + 6O_2 \longrightarrow 6CO_2 + 6H_2O$$

Glucose + oxygen $\longrightarrow$ Carbon dioxide + water + ENERGY

$$\swarrow \quad\quad \downarrow$$

ATP   Heat
for muscle
contraction

(ii) A: producers are the photosynthetic organisms, which trap sunlight energy, namely filamentous algae and diatoms.

B: primary consumers are the herbivores, which feed directly on plants or the producers, namely midge and blackfly larvae, mayfly nymphs and amphipod crustaceans.

C: secondary consumers are the carnivores, which feed on the primary consumers, namely leeches, stonefly nymphs and caddis larvae.

(iii) Young trout are omnivores feeding on plant producers (algae), and animal primary consumers (midge larvae and mayfly nymphs). Mature trout are carnivores, eating mainly secondary consumers (animals), e.g. leeches, stonefly nymphs, caddis larvae, and amphipod crustaceans.

(iv) The mayfly nymphs form an important *trophic level* in the food web. If they are removed, or killed by disease, the homeostatic balance within the food web will be affected. The caddis larvae will *decrease*, the diatoms will *increase*, and the number of young trout will be *reduced*.

(b) (i) Cold blooded means the body temperature fluctuates with the temperature of the external environment, since these animals are unable to regulate their body temperature.

(ii) Metabolic rate of the trout will increase up to the optimum temperature, above which the rate will fall and enzymes will be destroyed.

(iii) Solubility of gases and oxygen decreases with rising temperature.

(iv) Increasing water temperature would cause an increase in the metabolic rate and an increased demand for oxygen for aerobic respiration.

(v) Availability of oxygen will decrease as water temperature rises.

(vi) The trout will be deprived of oxygen and, being unable to release energy by aerobic respiration, will die. High temperatures (above $40°C$) can cause death through destruction of protein enzymes.

15 Basic Experimental Skills

15.1 Introduction

Practical work is an essential part of the GCSE Biology course and it involves certain basic experimental skills in order to carry out the practical investigations and show what you can do.

1. Hands are needed to perform *manipulative* skills (*manus*: Latin, meaning 'hand'), to handle pens, pencils, tools, specimens and apparatus.
2. Eyes, ears, nose and tongue are for *observation* to detect changes.
3. Writing skill is needed to record what you have *seen* and *done*.
4. Intelligent thought or reasoning is needed to *work out* or *interpret* the meaning of what you have seen happen in the experimental information or data, and may involve arithmetic *calculation* or use of *graphs*. Similarly thought or reasoning will be used in the *design* of new or improved experiments, or in order to draw *conclusions* from experimental investigation.

All these experimental skills can be summarised as follows, and many of these skills will be used many times throughout the course in different experimental investigations.

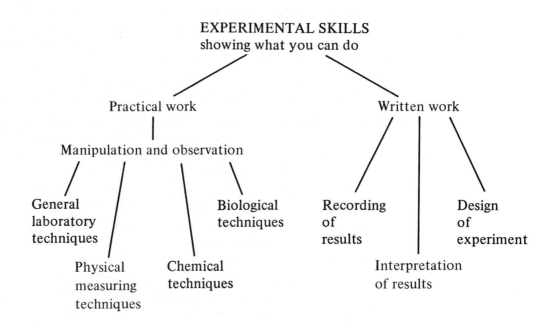

The *written* examinations will show the examiners what you *know and understand*, whereas the *practical* assessment or examination will show what you can *do*.

Students at full-time school or college will have their different practical skills assessed during the course by the teacher, and a score mark and date of achievement will be entered on a record sheet, or experimental skills card. External candidates will attempt a separate practical examination lasting about 1½ hours.

In the following pages details will be given as to how to acquire these different skills or techniques, and a space will be left for the mark awarded and the achievement date to be entered alongside. This will be a useful record for those working alone, or will supplement your teacher's assessment record.

15.2 General Laboratory Techniques

A laboratory is a work-room and can be a dangerous place with *hazards* from organisms, chemicals, equipment, apparatus, machines, gas and electricity.

(a) Personal Protection and Hygiene

1. For your own personal protection you will wear a cotton laboratory *coat* correctly buttoned up or a disposable polythene *apron*. Also you will need to know where protective *eyeshields, gloves, dust face-masks* and *safety screens* are located.
2. Personal hygiene means you *wash your hands* before and after experimental work. You must *not* eat or drink *anything*, and smoking is *forbidden* in the laboratory.

Mark . . . .　　　　　Achievement date . . . .

(b) Laboratory Protection and Hygiene

1. *Fire-drill* instructions must be known and understood.
2. *Unsupervised work* is forbidden and a teacher must be present at *all* times.
3. *First aid kit* position must be known.
4. *Spillages and breakages* must be reported immediately and the location of a special refuse disposal container must be known, and there must be an understanding of approved *first aid emergency treatments*.
5. *Water and gas taps* and electric switches must be turned off after use.
6. *Bench tops* should be wiped down after use with disinfectant solution.
7. *Pieces of equipment* must be returned clean and washed to their *correct* locations.

Do not interfere with experimental work that does not concern you, and leave the laboratory tidy as you found it!

Mark . . . .　　　　　Achievement date . . . .

(c) General Laboratory Equipment

1. *Glassware* is of two main kinds:
   (i) soft or *soda* glass, which is cheap and melts easily and may crack if heated suddenly;
   (ii) hard, or *boro-silicate* 'Pyrex' glass, which is more expensive and does not melt easily or crack on heating.
   Glassware includes: glass rods, tubing, test tubes, beakers, flasks, filter funnels, watch glasses, specimen bottles and petri dishes. Wash after use in warm deter-

gent solution, rinse and dry, or leave to air dry. Broken glassware must be picked up wearing *gloves*.

2. *Hardware* is mainly iron or wood apparatus, namely: *iron* tripod supports, retort stands, clamps and boss heads, wire gauzes. *Wood* is used for filter-funnel stands and test-tube racks.

   Always allow *hot ironware* to *cool* before handling. Recognise and know the names of all general laboratory equipment.

Mark . . . .   Achievement date . . . .

(d) Heaters and Heating Techniques

Heat is a form of *energy* needed for chemical and biological processes.

1. *Gas burners* can be of three main kinds: natural gas burners; bottled LPG (liquid petroleum gas) burners; and coal gas burners. Check the type in the laboratory.

   (i) *Connect the burner* correctly to the gas tap, pressing the connecting gas tubing firmly over the burner inlet and gas-tap outlet to prevent leakage of gas.
   (ii) *Light the gas* correctly and quickly.
   (iii) Regulate the height of the gas flame, and regulate the air inlet to produce a *quiet* burning flame of medium heat.

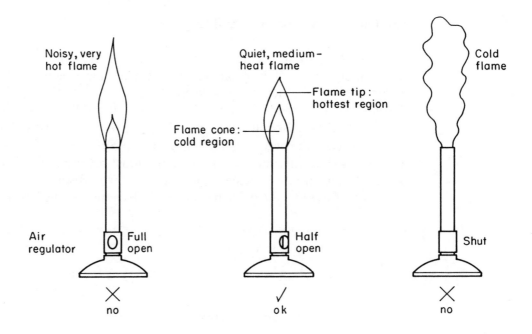

Figure 15.1 Gas burner technique.

2. Small ethanol *spirit* burners consisting of a glass container and wick can be used where there is no gas supply.

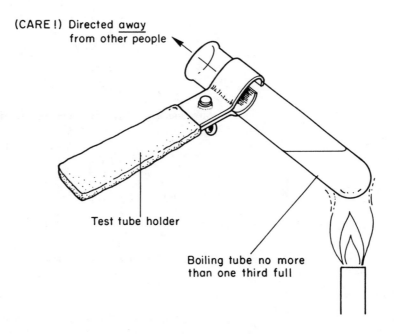

(CARE!) Directed <u>away</u> from other people

Test tube holder

Boiling tube no more than one third full

Figure 15.2 Boiling liquids in a boiling tube.

Mark . . . .   Achievement date . . . .

3. *Gas-burner heating technique*. Gas burners are sources of *direct heat* which can be applied directly to *hard glass* ('Pyrex') test tubes which must be supported by a *test-tube holder*. The test tube should *not* be more than one-third full of water or solution and its open mouth must be directed *away* from yourself or other people. The test tube should be shaken throughout the heating process.

There must be a *wire gauze* between a glass beaker or flat-bottomed flask and the iron supporting *tripod* stand; beakers and flasks must *not* be heated with a direct gas burner flame. Contents of beakers — water and aqueous solutions — should be stirred gently with a *glass rod* during heating. Wear protective *eye shields* to prevent hot liquid splashing outwards from the beaker while it is heating or boiling.

Only water and aqueous solutions of substances (i.e. substances dissolved in water) can be heated by these methods. Ethanol, which is flammable, *cannot* be heated in these ways.

Mark . . . .   Achievement date . . . .

4. *Water-bath heating technique*. Small amounts of flammable liquid such as ethanol must always be heated *indirectly* in a water-bath heater.

Biological processes require *heat energy*, and function best at around 35–40°C. This is achieved by using a bath of warm water, the temperature of which is carefully noted with a 0–110°C thermometer. The water bath can be a beaker, a metal pan or a small saucepan, heated by gas or electrically.

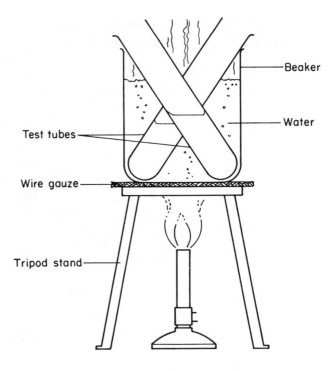

Figure 15.3 A simple water heating bath.

If the water-bath temperature rises too rapidly, it can be cooled down by adding small amounts of cold water. A student should be able to maintain the water-bath temperature within 2°C of the working temperature, i.e. 38 or 42°C for a 40°C working temperature. Special gas-regulating *thermostats* are inserted in the water bath to control the bath temperature automatically.

Mark . . . .   Achievement date . . . .

15.3 Physical Measurement Techniques

(a) Variables

Experimental investigation is concerned with observing *changes* or *variable* properties of substances. These variables are of two main kinds (see Sections 1.1(a) and 13.1(d)):

1. *Qualitative* variables *cannot* be measured by simple means, and include *colour, smell, taste, temperature*, and moisture (*humidity*). These variable properties can only be *compared* or *matched* by observation.
2. *Quantitative* variables *can* be measured as number, length, mass (weight), heat energy, pressure or time.

(b) Qualitative Observation

1. *Colour* can change, and the new shade or colour must be carefully noted and compared with *colour charts*. (Certain people may be colour-blind and unable to recognise reds and greens.) (See Section 1.2(c).)

   Hazard. Ultra-violet rays and infra-red rays from special lamps can damage the eyes. Therefore, always wear dark protective *goggles* and never look directly at these lamps.

2. *Smell* and various odours can be difficult to detect as this sense is more developed in some people than in others.

   Hazard. Do not smell substances without advice since they may be poisonous or produce *irritant* dusts or vapours (see Section 15.4).

3. *Taste.* The tasting of substances should not be done in any circumstances without advice.

   Hazard. This precaution is necessary since many substances are poisonous, harmful or corrosive.

4. *Temperature* is a variable change in hotness or coldness of a substance compared with either freezing ice or boiling water. It can be recorded on a scale called the Celsius ($^{\circ}$C) scale.

   The most common *thermometers* are the:

   (i) *laboratory* mercury-in-glass thermometer; ranges -10 to 50°C in 0.5°C graduation, -10 to 110°C in 1°C graduation, and -10 to 360°C in 2°C graduations;

   (ii) *clinical* thermometer, mercury in glass; range 35 to 42°C in 0.25°C graduations;

   (iii) *maximum and minimum* thermometer to record highest and lowest temperatures by metal markers reset with a magnet. Range -10 to 110°C in 1°C graduations.

   Hazard. Glass breaks easily, and mercury is *toxic*. Also note that clinical thermometers must be disinfected after use.

5. *Humidity* is the relative amount of moisture in the air, and is recorded as a relative percentage value by means of either a paper or hair *hygrometer*. Paper and hair are *hygroscopic* substances able to attract moisture or water vapour. *Normal* relative humidity values are between 45 and 75%. *Dry* is 20–30%, *wet* is 80–100%.

Mark . . . .   Achievement date . . . .

(b) Quantitative Measurement

Quantitative measurement involves the use of various *measuring instruments*.

1. *Number.* Counting of numbers of objects or organisms is done as follows. Small objects, such as daisy flower parts or different coloured pea seeds, should be placed on a dark-coloured background, black paper or black cloth; the latter *holds* the items better than paper, and the dark backgrounds help to show up the light-coloured object. Count in fives making 'five-barred gates' as follows: ⊥⊦⊦⊤ ⊥⊦⊦⊤ ⊥⊦⊦⊤. Thus three 'gates' equal 15 in total, and so on. Each stroke represents *one* item, and each gate is five items.

   Tally counters with push buttons are used to count large numbers of objects.

Marks . . . .   Achievement date . . . .

2. *Length* is measured by means of a metre (m) *rule* (100 cm in length) which has every centimetre divided into ten millimetres (mm). When taking a reading it is essential that the eye is *vertically over* the mark to be read and not in any side position which would cause a reading error (see Fig. 15.4).

   Callipers are used to measure thickness, or the inside and outside diameters of objects. The open calliper gap is then measured against a metre rule.

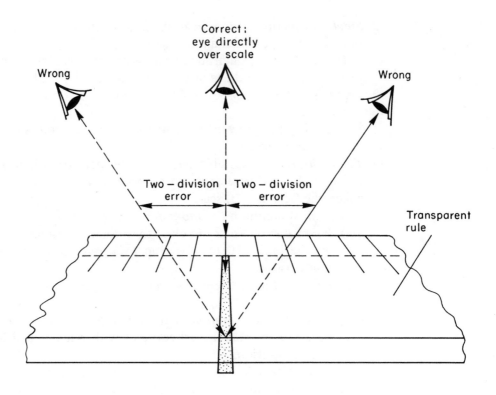

Correct:
eye directly
over scale

Wrong

Wrong

Two – division
error

Two – division
error

Transparent
rule

Figure 15.4.

Every GCSE candidate should be able to measure the length of objects correctly to within a millimetre (1 mm).

Mark . . . .   Achievement date . . . .

3. *Area* is measured in square metres, m², square centimetres, cm², or square millimetres, mm².

Most biological specimens have an *irregular* form, and their *surface area* is difficult to measure exactly.

For the *area-by-weighing method*:

(i) The surface area of a *leaf* is found by tracing around the *outline* on to thick card, then cutting out the outline which is carefully *weighed* as *a* grams.

(ii) A *square* of the *same* card is measured and its area is calculated by multiplying the square's *length* by its *width*. The square of card is weighed carefully (*b* grams).

(iii) The *leaf surface area* will be found as follows:
square card weighs *b* grams and has area *x* cm².
Leaf outline weighs *a* grams and its surface area will be (*a* grams × square area *x* cm²)/*b* grams, square card weight = *y* cm² or area of *one* leaf surface. Total leaf surface area = *y* cm² × 2.

For the *squared-graph-paper method*, graph paper is made up of one-millimetre (1 mm) squares (1 mm²) and larger, one-centimetre (1 cm) squares (1 cm²).

(i) A leaf can be placed on the graph paper and its outline drawn on to the paper.

(ii) The number of large (1 cm²) squares within the leaf tracing is counted, then every small (1 mm²) square that does not form part of a large cm square is counted by the 'gate' method.

(iii) The leaf area will be calculated as follows:

Total number of large (1 cm^2) squares = 32 cm^2

Total number of small (1 mm^2) squares outside
of large squares = 725 mm^2

Total area of *one* leaf surface = 32 cm^2 + (725/100)

= 32 + 7.25

= 39.25 cm^2

Two leaf surfaces = 39.25 × 2 = 78.5 cm^2

Mark . . . .   Achievement date . . . .

4. *Volume* is mainly measured by means of a graduated *measuring cylinder* or *graduated cylinders*. They are available as: 10 cm^3 capacity in 0.2 cm^3 graduations, 50 cm^3 capacity in 1 cm^3 graduations, and up to 1000 cm^3 capacity in 10 cm^3 graduations.

   Reading the liquid level is an important technique in which the eye must be level with the *lower* liquid level of the *meniscus* as shown in Fig. 15.5, otherwise reading errors will occur.

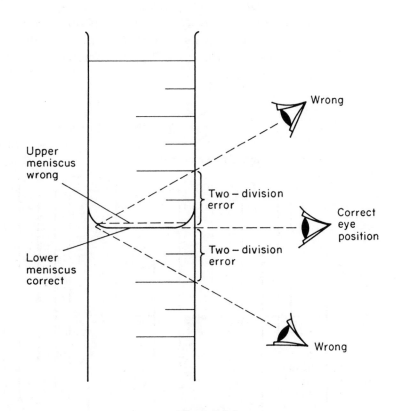

Figure 15.5.

Non-living biological specimens can be lowered into a *half*-filled measuring cylinder of water until completely submerged. The increase in water level is noted as *b* cm^3. The volume of the object will then be obtained by *subtracting* the reading for the half-filled cylinder *a* cm^3 from *b* cm^3.

Mark . . . .   Achievement date . . . .

5. *Mass or weight* can be determined by *spring balances* with a capacity of 100 g graduated in 1 g divisions, the specimen being supported on a small pan attached to the balance hook. Larger-capacity spring balances weighing up to 1000 g (1 kg) are available. A *simple lever balance*, with a range up to 1000 g or 250 g in 1 g divisions, will weigh objects directly with reasonable accuracy. *Double*

pan balances require a box of weights, and they weigh to an accuracy of 10 mg or 0.01 g.

Every GCSE candidate should be able to weigh an object to an accuracy of 1 g; i.e. if the accurate weight of a seed potato is 52 g, then a value of either 51 or 53 g is acceptable.

Mark . . . .   Achievement date . . . .

6. *Time* is recorded by means of a stop watch or the student's own personal watch.

7. *Estimating* is an important technique which candidates should undertake in estimating or making reasonable guesses for length, weight (mass), and volume of a single object. With skill the estimate can be within 10% of the correct value.

Mark . . . .   Achievement date . . . .

8. *Pressure* changes are indicated by the *manometer gauge*. This is a U-shaped glass tube containing coloured water (see Fig. 15.6).
    (i) *Equal* levels in the manometer gauge show pressure inside the apparatus is *equal* to outside air pressure.
    (ii) *Falling* level towards the apparatus shows pressure outside the apparatus is *greater* than that inside.
    (iii) *Rising* level away from the apparatus shows increasing pressure inside the apparatus *greater* than surrounding outside air pressure.

Differences in levels in the manometer are measured carefully with a rule in centimetres (cm).

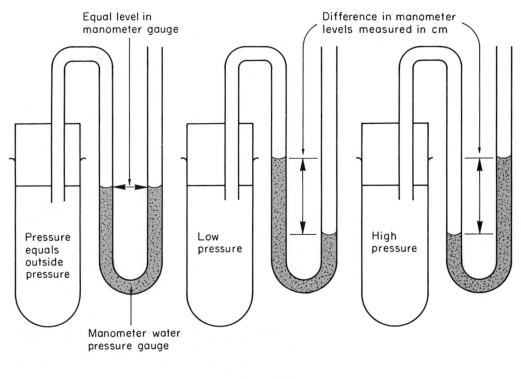

Figure 15.6.

9. *Energy* contained within a substance, e.g. food, or a mammal is measured in joules (J), kilojoules (kJ), and megajoules (MJ). It depends on the body's *mass* (weight) as well as its *temperature*.

Various methods are used to measure energy using *calorimeter* apparatus.

15.4 Chemical Techniques

(a) Chemical Reagents

Chemical reagents must be used with *care* and under *supervision*. Some are potentially dangerous or hazardous and can cause fire, explosion, or death by poisoning, or are harmful to the body as irritants, or are corrosive. *Hazard symbols* will be seen on the reagent bottle label, as indicated in Fig. 15.7.

Table 15.1 shows some chemical reagents you may expect to use, together with their hazards.

Table 15.1   Chemical reagents and their hazards

Chemical reagent	Hazards in use
Acetic acid (ethanoic acid)	*Corrosive*, vapour *irritates* eyes and lungs. Burns skin and eyes
Alkaline pyrogallol	*Corrosive*, causes skin and eye burns. *Poisonous* if swallowed and absorbed by the skin
Ammonium hydroxide or ammonia solution	Vapour *irritates* lungs. Solution corrosive to skin and eyes, *poisonous* if swallowed
Copper sulphate and Fehling's solution	*Irritates* eyes; if swallowed causes vomiting and diarrhoea
Calcium chloride	*Irritant* dust and highly hygroscopic
Ethanol	*Flammable, poisonous* if swallowed, may cause blindness
Formaldehyde (methanal)	*Irritant* vapour affects lungs and eyes. *Poisonous* if swallowed
Hydrochloric acid	Vapour *irritates* eyes, skin and lungs. Liquid *corrosive* and *poisonous*
Hydrogen peroxide	*Oxidiser* supports burning. Liquid *irritates* eyes and skin. Causes vomiting and internal bleeding if swallowed
Iodine solution	Vapour *irritates* eyes and lungs. Highly *poisonous* if swallowed
Soda lime	Causes skin and eye burns, *irritant* dust affects lungs. Harmful if swallowed
Sodium and potassium hydroxides	*Corrosive*. Solutions burn the skin and eyes. Severe damage caused if swallowed
Propanone (acetone; nail varnish remover)	*Flammable*, poisonous if inhaled or swallowed

Solid chemical reagents must always be handled by means of special metal *spatulas* or *spoons*, placing the reagent on clean glass dishes or *watch glasses*. NEVER touch, taste or smell any chemical reagent. Droppers are used to transfer small amounts of *liquid* chemical reagents.

At this point every GCSE candidate should know and recognise the hazardous chemicals they may encounter.

Mark . . . .   Achievement date . . . .

(b) Separation Techniques

Distilled or *deionised* water (*not tap water*) is the main *solvent* used in separation of chemical substances. *Ethanol* (flammable) is used to a lesser extent. The following methods refer to the use of water as the solvent.

Figure 15.7 Hazard symbols.

1. *Filtration* involves mixing and warming the mixture to dissolve soluble components. The *insoluble residue* is collected by pouring the mixture through a *filter paper* shaped into a cone by folding twice. The moistened filter paper is held in the glass *filter funnel*. The clear solution which passes through the filter paper is called the *filtrate*.

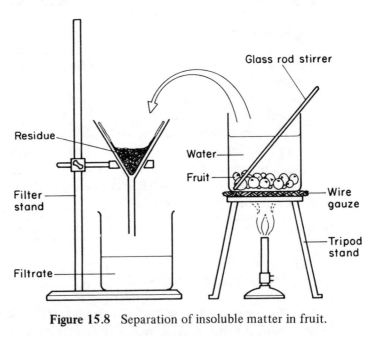

Figure 15.8 Separation of insoluble matter in fruit.

Mark . . . .   Achievement date . . . .

276

2. *Evaporation* is the separation process in which the water *solvent* is evaporated to leave a solid *residue*. The filtrate or clear solution is placed in a porcelain evaporating basin and heated to dryness either directly over a bunsen burner or indirectly over a water bath.

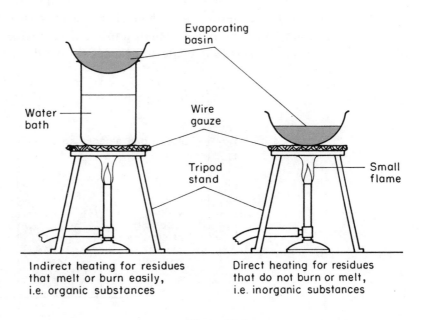

Figure 15.9.

Mark . . . .   Achievement date . . . .

Desiccators are used to keep substances that have been heated to dryness in a *dry* condition. Substances need to be heated until they show a *constant* or steady unchanging weight proving that all water has been removed. This is important in soil analysis (see Section 14.2). The substance is placed over *anhydrous* calcium chloride which is very *hygroscopic*. Figure 15.10 shows a simple desiccator.

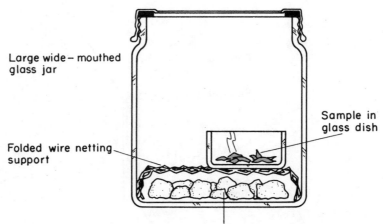

Figure 15.10 Desiccator for general use.

Mark . . . .   Achievement date . . . .

3. *Dialysis* is the separation of substances made up of small molecules, e.g. mineral ions, glucose and urea (carbamide), from large-molecule substances such as proteins and the polysaccharide, starch.

A special *semi-permeable* membrane called *dialysis tubing* (visking) is used. This tube has a knot tied in it at one end, and the tube is filled with the solution to be separated, and then closed with a knot tied in the open end. The 'sausage' thus produced is then immersed in distilled or deionised water which receives the small-molecule substances (mineral ions, glucose, etc.) by dialysis.

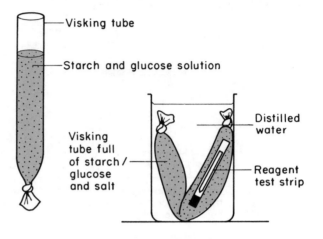

Figure 15.11.

4. *Chromatography* is a method for separating *coloured* chemical compounds.

The individual *chlorophyll* pigments are extracted from dried nettle leaves by grinding with a small amount of *propanone* (acetone) (Care! *Flammable*: observe fire precautions. This experiment should be demonstrated by a supervisor.)

The grinding process is performed in a *mortar* and *pestle* to extract the pigments that dissolve in the solvent. The solvent extract is filtered through glass wool.

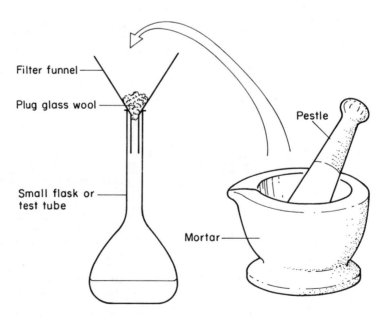

Figure 15.12 Mortar and pestle used for pigment extraction.

The extract is then placed as a central spot on the filter paper supported over a *petri dish* containing propanone (acetone) solvent as in Fig. 15.13. This apparatus must be placed aside and away from naked flames, and fire precautions must be observed.

In time the solvent rises up the 'tail' and causes the pigments to separate into different bands of colour.

Bright green   – chlorophyll A
Brownish yellow   – xanthophyll
Blue–green   – chlorophyll B
Yellow   – carotenes

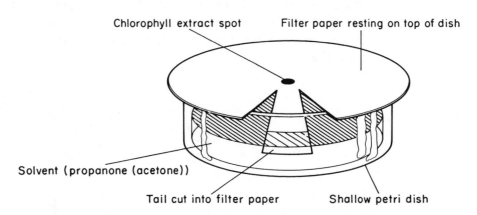

Figure 15.13 Separation of chlorophyll pigments by paper chromatography.

Mark . . . .   Achievement date . . . .

5. *Solution concentration* is generally indicated as a percentage weight (mass) to volume or % w/v. A known weight (mass) of substance is dissolved in 100 cm³ of water. For example, 5 g of substance or *solute* is dissolved in 100 cm³ of water to make a 5% w/v solution.

A 0.5% solution could be made by dissolving 5 g in 1000 cm³ of water; this is more convenient than trying to weigh out 0.5 g of a substance.

Different solution strengths of common salt from 0.5, 1, 2 to 10% should be prepared using a simple balance and measuring cylinder. The solutions should be placed in *labelled* bottles.

Mark . . . .   Achievement date . . . .

6. *Other methods* of separation include *distillation*, and the use of *centrifuges* which separate *insoluble* substances in solutions by a fast spinning process.

Mark . . . .   Achievement date . . . .

15.5   Biological Techniques

(a) Handling Organisms

Living organisms, whether large (macro-organisms) or microscopically small (micro-organisms), must at all times be treated with due care, since many wild animals and micro-organisms are sources of *disease* and *infection*.

1. *Wild* pigeons may cause psittacosis, a lung disease which can be fatal; *wild* rats may cause Weil's disease, also potentially fatal; and other mammals, birds and reptiles can cause salmonella food poisoning. *Strong gloves* should be worn, and any bites or scratches must be treated immediately by a doctor.

2. *Micro-organism cultures* (Section 14.2 and Fig. 14.7) must be handled with the greatest of care, and always with the permission and under the strict supervision of your instructor.

   (i) *Do not remove* culture dish lids: they must be kept taped down; and do not unscrew caps of culture bottles.

   (ii) *Dispose* of micro-organism cultures by soaking in strong disinfectant or by burning in a furnace.

   Wash your hands thoroughly after handling any living organism or its container.

   Metal or plastic *forceps* with *blunt* points are used to pick up large objects such as seeds, and forceps with *fine* points are used to pick up flower parts. *Pooters* are used to pick up small insects. Small objects in water can be picked up by means of either a *camel-hair brush*, or a *dropper pipette* fitted with a teat (see Fig. 15.16).

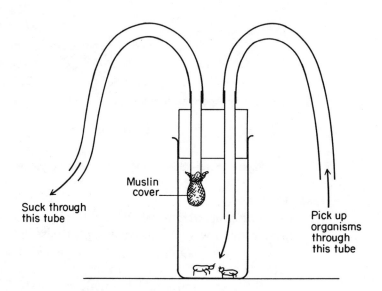

Figure 15.14 A pooter collecting tube.

(b) Dissection

This is a means of showing up structures in plants or animals. Sharp *fine-pointed scissors* are used to open, for example previously stained celery stems. *Scalpels* are sharp knife-like instruments used to cut through stem tissues or fruits, aided by *needles* and *seekers*, which are used to part the tissues.

Dissection instruments must be *washed* in hot soapy water and disinfectant after use, and kept in a cloth instrument roll.

(c) Magnifiers and Microscopes

Magnifiers and microscopes *magnify* the size of an object's *image*. The degree of magnification is shown by the symbol × followed by the number of times by which the image is magnified.

1. *Magnifiers* come in a range of magnifications, × 5, × 10 and × 20. Most *small objects* such as certain insects or flower parts are examined with a magnifier, also called a *hand lens*. GCSE students should practise drawing objects using a magnifier.

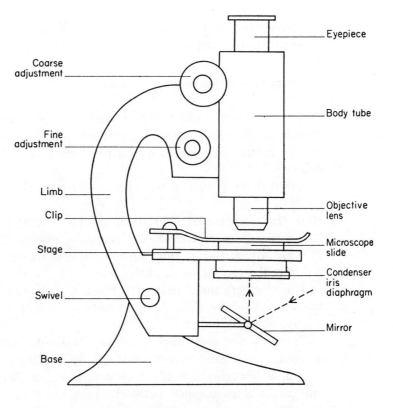

Figure 15.15 Parts of a light (optical) microscope.

2. *Microscopes* (Fig. 15.15) are means of magnifying an object using *two* lenses: the *eyepiece* lens, and the *objective* lens.

The simplest microscope for GCSE examination use has an eyepiece with × 10 magnification, and *objective lenses* giving × 4, × 10 or × 20 magnification. The magnifications obtained are as follows:

Eyepiece		Objective	
× 10	× 4	× 10	× 20
Magnification	40 times	100 times	200 times

1. *Microscope Care*

Microscopes are valuable biological instruments and are expensive to buy. Lenses should *not* be touched with the fingers, and should be dusted with a clean camel-hair brush. Always keep the microscope under its cover when not in use. Soft paper tissues are used to dry a wet lens.

2. *Microscope Use*

(i) Light from a table lamp (*not* sunlight) is directed on to the mirror, and adjustment of the diaphragm is made to vary the illumination to suit you and the object you are looking at.

(ii) Select the low-power × 4 objective lens, and with your eye level with the microscope stage slowly rack the objective *downwards* until you see it nearly touch the microscope slide.

(iii) With eye at the eyepiece *slowly* rack the microscope tube *upwards* until the image comes into view.

(iv) Change to the *high-power* × 10 or × 20 objective lenses and adjust the illumination to brighten the object.

(d) Preparation of Microscope Material

1. *Cavity microscope slides* have a small hollow into which a drop of water containing small organisms is placed from the dropper pipette. This slide is placed on the microscope stage, which must be in the horizontal and not the tilted position.

    After viewing, return the specimen by gently rinsing it off into its original container.
2. *Plain flat microscope slides* are used for preparing temporary mounts of sectioned or sliced specimens of, for example, potato sliced with a razor (see Fig. 15.16). Alternatively, thin strips of lining tissue, or epidermis removed from onion bulbs, are mounted in this way.

    The *thin* specimen tissue is picked up by means of a camel-hair brush and transferred to the slide centre position. It may be necessary to spread it out *gently* with needles.

    Carefully add *one* drop of iodine solution to stain the tissue. Carefully lower a square cover glass by one edge, gently guiding it into position with a needle. Continue lowering the cover glass until it lies flat over the specimen without any air bubbles trapped between the glass. If air bubbles are present, discard the specimen and repeat the preparation until *no* air bubbles are present. Any surplus liquid on the slide surface must be mopped up by means of blotting paper.

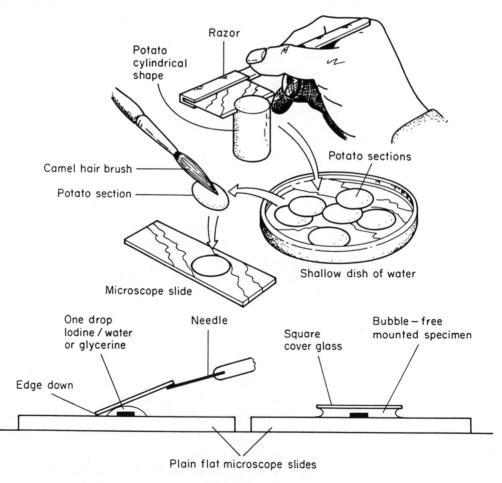

Figure 15.16.

282

Mark . . . .   Achievement date . . . .

Photographs of specimens taken through a light microscope are called *photo-micrographs* and are available as prints of 35 mm transparencies. The electron microscope provides photographic prints called *electron micrographs* which show cell structures magnified as much as 500 000 times.

(e) Sterilisation of instruments and glassware

Sterilisation of equipment is often necessary and this is done by heating in a *pressure cooker* or *autoclave* with water for 20 minutes. Glassware and metalware can be sterilised by *flaming* or passing the article through a gas burner or spirit-lamp flame. Skin is sterilised by wiping with cotton wool soaked in 70% ethanol solution.

Mark . . . .   Achievement date . . . .

15.6   Written Work

Written work is the equally important part of experimental investigation, and should be kept in a special practical notebook, ring file or folio.

(a) Diagrams

Diagrams are essential records for describing specimens or laboratory apparatus; they are in themselves descriptions and do not need a written description in prose or words, apart from the labels that every diagram must have.

Similarities or differences between specimens can be described in diagrams by drawing attention to these points by means of prominent arrows.

1. *Line diagrams* are drawn with a *single* line showing the clear outline shape of a specimen, using a sharp pencil. They are drawn in *two* dimensions, and are not supposed to be artistic sketches showing perspective or three dimensions, or background scenery, etc.

    Diagrams of *laboratory* apparatus should be drawn with a *ruler*, and *stencils* may be used for apparatus outlines.

2. *Plan diagrams* can be compared to 'maps' showing areas where different tissues are found. This type of diagram is used for plant stem, root and leaf structure (see Figs 3.2 and 4.1).

3. *Flow diagrams* are also useful for describing the *sequences* or stages in a *process* involving more than one stage or technique, and replace written descriptions (see Fig. 15.18).

    Labelling of all diagrams must be in *ink* with clear pointers or lines connecting the label with the named part of the apparatus or organism.

    Scale size should be indicated on the diagram for macro-organisms in centimetres, and the approximate *magnification* should be indicated for micro-organisms.

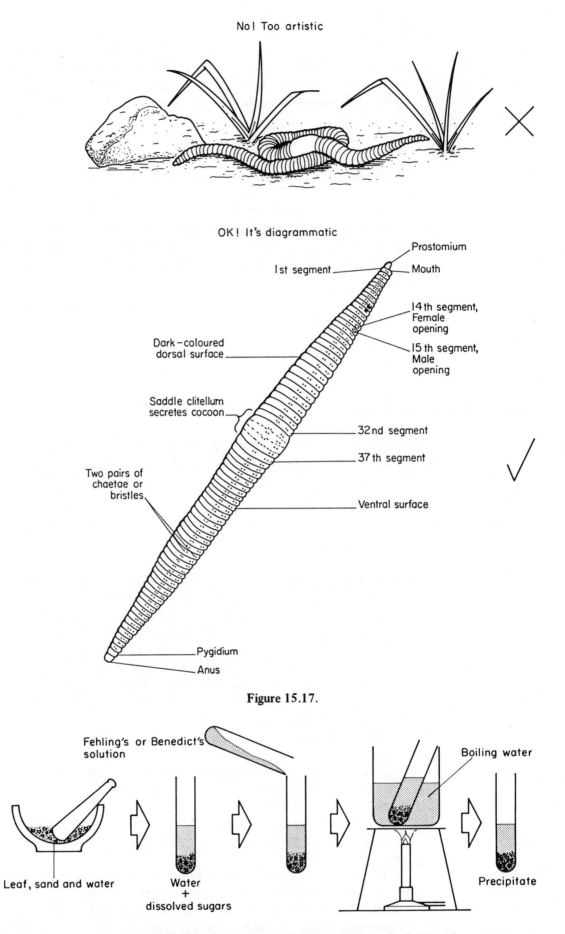

No! Too artistic

OK! It's diagrammatic

Prostomium

1st segment ——— Mouth

14th segment,
Female
opening

Dark—coloured
dorsal surface

15th segment,
Male
opening

Saddle clitellum
secretes cocoon

32nd segment

37th segment

Two pairs of
chaetae or
bristles

Ventral surface

Pygidium

Anus

Figure 15.17.

Fehling's or Benedict's
solution

Boiling water

Leaf, sand and water

Water
+
dissolved sugars

Precipitate

Figure 15.18 Flow diagram for sugar testing in leaf material.

284

(b) Information Recording

Every experimental investigation produces facts or *information* which may be *qualitative* (concerned with the colour, shape or form, or chemical composition of a substance or organism); or it may provide *quantitative* information in *figures* or values of weight, length or pressure. This information must be recorded neatly in table form, and figures and numbers must be written correctly showing decimal points, followed by correct symbols for measuring units: °C, g, mg, cm, cm² or cm³, etc. For example, if a seed weighs (has a mass of) three-and-a-half grams, it is not enough to write down 3.5: the symbol for the measuring unit, g, must be included; thus 3.5 g is correct. Data or information is *incomplete* without mentioning the unit of measurement.

For example, if a process is known to have taken 12 units of time from start to finish, it must be specified whether the time taken was 12 s, 12 m, 12 h or 12 d, namely seconds, minutes, hours or days.

(c) Numerical Information Display

Neat tabular records of experimental information in the form of *numerical data* require some form of presentation or display. This can be as:
1. *calculations* of numerical relationships; ratios, percentages or averages;
2. *pictorial* or graphical display as pie diagrams, bar charts or histograms and line graphs.

The following data refer to the *numbers* of certain plants in a crop having a certain *height*.

(i) In this experimental investigation, practical measurement of the numbers of plants was by the 'gate' or tally count method, and height was recorded in centimetres with a 2 metre rule.
(ii) The data were arranged in a neat tabular way showing the unit of height measurement, cm. The data in Table 15.2 form the basis of several calculations in the following pages.

Table 15.2   Tabular record of number of plants and their height

Number of plants	Height (cm)
2	55
3	60
5	65
10	70
13	75
9	80
25	85
30	90
34	95
41	100
56	105
32	110
22	115
8	120
3	125

(a) *Calculations*

1. *Average Values*

Average values are calculated by *adding* together *all* the plant number or hight values and *dividing* by the number of groups or sets. You will see in Table 15.2 that there are 15 sets or groups of plant heights and plant numbers. To calculate the *average number of plants* in a group:

1. Add all the plant numbers together in the results table from 2 + 3 + 5 . . . to 22 + 8 + 3 using a calculator. Total = 293.
2. Divide this by total number of sets, i.e. 15. Therefore: average number of plants in each group = 293 ÷ 15 = 19.53.

To determine the *average height of all plants* in the experimental investigation:

1. Add together all the plant height values in Table 15.2 from 55 cm + 60 cm . . . to 120 cm + 125 cm. Using the calculator this totals as 1350 cm.
2. Divide this total value by the number of sets, i.e. 15. Therefore: average plant height = 1350 ÷ 15 = 90.0 cm.

2. *Ratios*

A ratio is the numerical relation that one quantity bears to another. The investigators whose results are shown in Table 15.2 grouped all plants of 85 cm and under as *dwarf* plants and the remainder were called *tall* (over 85 cm in height). Table 15.2 lists 67 as *dwarf* (85 cm or under in height), and 293 − 67 = 226 are *tall* and over 85 cm in height.

Ratios are calculated by dividing the *smallest* number into the *largest* number.

$$\text{Ratio of tall to dwarf plants} = \frac{226 \text{ tall}}{67 \text{ dwarf}} = 3.4 \text{ to } 1$$

This is also written with the symbol ':' as shown:

Ratio of tall to dwarf plants = 3.4 : 1

3. *Percentages*

Percentage value is a part of one hundred expressed in hundredths. Percentages are calculated from the following relationship:

$$\frac{\text{Number of certain quantity}}{\text{Total number all quantities}} \times 100 = \%$$

For example, the percentage of dwarf plants in the investigation will be

$$\frac{67 \text{ dwarf}}{293, \text{ the total of all plants, dwarf and tall}} \times 100 = 22.87\%$$

or out of all the plants 22.87% are *dwarf* and 100% − 22.87% or 77.13% are *tall* plants.

(b) *Pictorial and Graphical Displays*

These are attractive visual methods of presenting experimental data.

1. *Pie Diagrams*

Pie diagrams are *circular* diagrams cut by radii into segments or *sectors* ('pie slices') illustrating relative magnitudes or frequencies. A circle is composed of 360°, and the angle of each sector can be measured with a protractor.

The calculations for pie diagrams are based on the following relationship:

$$\frac{\text{Individual quantity}}{\text{Total quantity}} = \frac{\text{Degrees in sector}}{360° \text{ in circle}}$$

Thus the total quantity fills the whole circle, and individual or component quantities occupy sectors.

The number of dwarf plants, i.e. 67 that are 85 cm or under in height, can be shown as a pie-diagram sector.

$$\frac{67}{293 \text{ (total quantity)}} = \frac{\text{Degrees in sector}}{360°}$$

$$\text{Degrees in sector} = \frac{360° \times 67}{293}$$

$$67 \text{ dwarf plants} = 82.3°$$

A large circle is drawn and the 82° sector is marked off to produce the complete pie diagram as shown in Fig. 15.19.

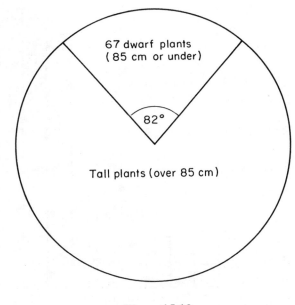

67 dwarf plants
(85 cm or under)

82°

Tall plants (over 85 cm)

Figure 15.19.

2. *Bar Charts*

Bar charts are means of showing the *distribution* or variation of the different plant heights by means of bars or rectangles. *All* the data of the experimental investigation can be displayed attractively in this way.

Graph or *squared paper* is needed for this display. The *steadily increasing* quantities, namely plant height figures, are arranged on the *horizontal* line or axis, and the changing or *fluctuating* values for plant numbers are arranged on the *vertical* line or axis, as shown in Fig. 15.20.

3. *Line Graphs*

Line graphs are diagrams that show relationships between *two* changing or variable quantities. The graph must be drawn on squared paper, and the *horizontal* line is

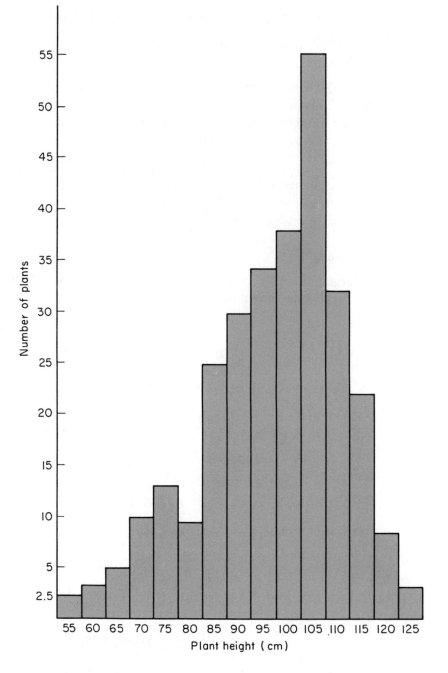

Figure 15.20 Bar diagram showing height distribution among plants.

used to record *steadily* increasing quantities, whereas the *vertical* line is for variable *fluctuating* quantities.

The data for distribution of plant height given earlier are not suitable for display on a line graph.

The following data concern changes in body weight of an insect over a time period of 24 days. Two variables are present, namely changing body mass (*weight*) and *time*.

Day	1	3	6	9	12	15	18	21	24
Body mass (weight), mg	100	100	200	230	350	530	600	760	1200

Since day (time) changes steadily (as an independent variable), we plot it on the horizontal line. Mass (weight) is plotted vertically since it shows fluctuation in the early data.

Each mass (weight) and day (time) value is marked neatly on the graph with a cross, and each plotted position or cross is joined by connecting lines as shown in Fig. 15.21.

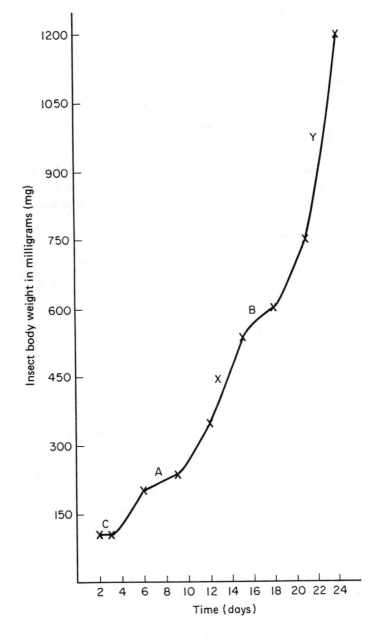

Figure 15.21.

Information can be obtained from looking at the *shape* of a graph. Most of the curve in Fig. 15.21 is steep, showing rapid mass (weight) increase, but two parts, A and B, are less steep, showing little mass (weight) increase, and part C is horizontal, showing no mass (weight) increase.

This difference can be calculated as a percentage mass (weight) increase for different parts of the graph.

The percentage mass (weight) increase for

A is $230 - 200 = \dfrac{30}{200} \times 100 = 15\%$

B is $600 - 530 = \dfrac{70}{530} \times 100 = 13\%$

C is $100 - 100 = 0\%$.

Other graph portions show the following large mass (weight) increases. Between days 12 and 15 it is:

$x = 530 - 350 = \dfrac{180}{350} \times 100 = 51\%$

Between days 21 and 24 it is:

$y = 1200 - 760 = \dfrac{440}{760} \times 100 = 58\%$

See questions 1.2, 7.3, 8.3, 9.2, 10.1, 14.1, 14.2 and 14.3 and Figs 1.3, 4.8, 7.1, 7.7 and 8.11, all of which show the importance of graph construction and interpretation in solving biological problems and data.

16 Complete Course Self-test

The following self-test is by means of *matching pairs* or by *classification*-type objective questions which are used by some examining groups (NEA, NISEC, L & EAG) in their *compulsory* basic-level written examinations.

16.1 Bioscience

Questions 1, 2, 3 and 4

    A. microgram
    B. micrometre
    C. kilogram
    D. litre

From the above list choose one which is used to

Q.1   record human body mass   Answer . . . .

Q.2   measure human lung volume   Answer . . . .

Q.3   measure daily intake of vitamins
          A and D in humans   Answer . . . .

Q.4   measure the diameter of a cell.   Answer . . . .

Questions 5, 6, 7 and 8

    A. lipid
    B. water
    C. air
    D. protein
    E. glucose

Which of the chemical substances listed above

Q.5   is a carbohydrate?   Answer . . . .

Q.6   is a mixture of chemical elements
          and compounds?   Answer . . . .

Q.7   contains the component element nitrogen?   Answer . . . .

Q.8   is composed of the chemical elements
          hydrogen and oxygen *only*?   Answer . . . .

Questions 9, 10, 11 and 12

    A. radiation
    B. diffusion

C. evaporation

D. ionisation

E. decomposition

Which of the above named processes will occur when

Q.9   salt (sodium chloride) dissolves in water?   Answer . . . .

Q.10   the Sun gives out light energy?   Answer . . . .

Q.11   the human body loses heat?   Answer . . . .

Q.12   water changes from a liquid into a
       vapour on hot skin?   Answer . . . .

Questions 13, 14, 15 and 16

A. 12 MJ

B. pH 7

C. 100 kPa

D. 37°C

E. 17 kJ/g

Which of the above values and symbols indicates

Q.13   the average body temperature?   Answer . . . .

Q.14   the average air pressure?   Answer . . . .

Q.15   the daily energy need of an adult man?   Answer . . . .

Q.16   that water is neutral to litmus?   Answer . . . .

16.2   Variety of Organisms

Questions 17, 18, 19 and 20

A. fungi

B. algae

C. mosses

D. ferns

E. angiosperms

Which of the plant groups listed above

Q.17   lacks chlorophyll?   Answer . . . .

Q.18   produces seeds?   Answer . . . .

Q.19   are decomposers?   Answer . . . .

Q.20   are almost all aquatic?   Answer . . . .

(NISEC)

Questions 21, 22, 23 and 24

A. molluscs

B. arthropods

C. annelids

D. amphibians

E. mammals

Which of the animal groups listed above

Q.21   has an exoskeleton, segmented bodies
       and jointed limbs?   Answer . . . .

Q.22   has metamerically segmented worm-like
       soft bodies?   Answer . . . .

Q.23   has non-segmented soft bodies with a
        muscular foot and chalky shell?   Answer . . . .
Q.24   has four-limbed bodies covered with
        hair, and in which the female feeds
        milk to its young?   Answer . . . .

Questions 25, 26, 27 and 28

        A. grass
        B. mushroom
        C. rabbit
        D. earthworm
        E. virus
Which of the above living organisms
Q.25   has chloroplasts in its cells?   Answer . . . .
Q.26   excretes urine containing nitrogenous
        waste?   Answer . . . .
Q.27   is autotrophic and a food producer?   Answer . . . .
Q.28   consists mainly of cellulose?   Answer . . . .

Questions 29, 30, 31 and 32

        A. pine trees
        B. spiders
        C. apple trees
        D. shrimps
        E. insects
Which of the above
Q.29   have two pairs of antennae?   Answer . . . .
Q.30   produce fruits?   Answer . . . .
Q.31   have only three pairs of legs?   Answer . . . .
Q.32   have four pairs of legs?   Answer . . . .

16.3 Cells, Tissues, Organs and Organisation

Questions 33, 34, 35 and 36

        A. cell walls
        B. mitochondrion
        C. cell membrane
        D. small vacuoles
        E. glycogen granules
Which of the above
Q.33   is an organelle of aerobic respiration?   Answer . . . .
Q.34   is never part of animal cells?   Answer . . . .
Q.35   is a fluid-filled compartment within cells?   Answer . . . .
Q.36   is called animal starch?   Answer . . . .

      A. epithelial tissue
      B. muscle tissue
      C. connective tissue
      D. nerve tissue
      E. glandular tissue

Which of the animal tissues listed above has a

Q.37   contractile function?   Answer . . . .

Q.38   protective lining and covering function?   Answer . . . .

Q.39   communication function?   Answer . . . .

Q.40   secretory function?   Answer . . . .

Questions 41, 42, 43 and 44

      A. xylem tissue
      B. chlorenchyma mesophyll tissue
      C. epidermal tissue
      D. parenchyma tissue
      E. meristem tissue

Which of the plant tissues listed above

Q.41   is concerned with food manufacture
       in leaves?   Answer . . . .

Q.42   is vascular tissue for transport of water
       and solutes?   Answer . . . .

Q.43   is mainly the component tissue of root
       and stem cortex?   Answer . . . .

Q.44   is the outermost layer of cells in a
       plant body?   Answer . . . .

Questions 45, 46, 47 and 48

      A. glycerol (propanetriol)
      B. flat-sided razor
      C. hand lens
      D. iodine solution
      E. ethanol

Which of the above is used to

Q.45   cut a transverse section of carrot tissue?   Answer . . . .

Q.46   stain the section?   Answer . . . .

Q.47   mount the section on a slide?   Answer . . . .

Q.48   harden and preserve soft specimens?   Answer . . . .

16.4   Plant Nutrition

Questions 49, 50, 51 and 52

The accompanying diagram illustrates an experiment to test a leaf disc for starch. Discs were transferred in the direction shown by the arrows.

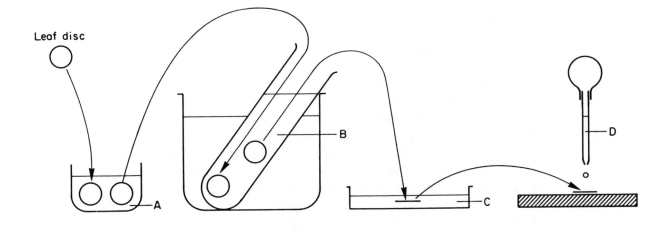

Leaf disc

Which of the liquids A to D is

Q.49   iodine in potassium iodide solution?   Answer . . . .
Q.50   ethanol (alcohol)?   Answer . . . .
Q.51   used to soften the discs?   Answer . . . .
Q.52   used to decolourise the discs?   Answer . . . .

(NEA)

Questions 53, 54, 55 and 56

    A. cuticle
    B. guard cells
    C. palisade mesophyll cells
    D. vascular bundle
    E. epidermis

Which of the above has the function of

Q.53   preventing excessive water loss due to
        its lipid composition?   Answer . . . .
Q.54   being the main region of photosynthesis? Answer . . . .
Q.55   protecting the underlying leaf tissue?   Answer . . . .
Q.56   opening and closing the stoma?   Answer . . . .

Questions 57, 58, 59 and 60

    A. oxygen
    B. carbon dioxide
    C. light
    D. water
    E. iron ions

Which of the above

Q.57   is present in air to the extent of 0.03%?   Answer . . . .
Q.58   provides hydrogen for carbohydrate
        synthesis?   Answer . . . .
Q.59   enters the plant root hairs by active uptake? Answer . . . .
Q.60   acts as a limiting factor when its intensity
        alters?   Answer . . . .

Questions 61, 62, 63 and 64

    A. calcium ions
    B. phosphate ions
    C. magnesium ions
    D. nitrate ions
    E. potassium ions

Which of the above is

Q.61   essential for the formation of the
        chlorophyll molecule?   Answer . . . .

Q.62   needed to form intercellular cell wall
        cement?   Answer . . . .

Q.63   essential for amino acid formation?   Answer . . . .

Q.64   needed for energy transfer and release?   Answer . . . .

16.5 Animal Nutrition

Questions 65, 66, 67 and 68

    A. amino acids
    B. glycerol (propanetriol)
    C. cellulose
    D. water
    E. glucose

Which of the above

Q.65   is the main reactant in chemical digestion
        of foods?   Answer . . . .

Q.66   is not a product of chemical digestion
        of foods?   Answer . . . .

Q.67   is the end product of protein digestion?   Answer . . . .

Q.68   has no energy value as a herbivore food
        component?   Answer . . . .

Questions 69, 70, 71 and 72

    A. iron ions
    B. vitamin C (ascorbic acid)
    C. calcium ions
    D. vitamin D (cholecalciferol)
    E. vitamin A (retinol)

Which of the above is

Q.69   a mineral requirement for blood formation? Answer . . . .

Q.70   a mineral requirement for teeth and bone
        formation?   Answer . . . .

Q.71   needed for calcium ion uptake?   Answer . . . .

Q.72   manufactured in the body from carotenes
        present in certain plant-source foods?   Answer . . . .

A. stomach
B. large intestine
C. dietary fibre
D. liver
E. small intestine

Which of the above
Q.73 provides bulk to aid peristalsis?   Answer . . . .
Q.74 secretes emulsifiers for lipid emulsification?   Answer . . . .
Q.75 is the part of the gut mainly concerned with
      food absorption and digestion?   Answer . . . .
Q.76 secretes a strongly acid (pH 1) digestive
      juice?   Answer . . . .

16.6   Transport

Questions 77, 78, 79 and 80

A. active transport
B. peristalsis
C. diffusion
D. evaporation
E. osmosis

Which of the terms listed above
Q.77 is a form of mass flow in tubular organs
      caused by muscular contraction?   Answer . . . .
Q.78 is a movement of materials against a concen-
      tration gradient using metabolic energy?   Answer . . . .
Q.79 is the change of a liquid into a vapour?   Answer . . . .
Q.80 is the movement of particles from a region
      of high concentration to a region of
      lower concentration?   Answer . . . .

Questions 81, 82, 83 and 84

A. xylem
B. water
C. phloem
D. radiant energy
E. mineral ions

Which of the above
Q.81 enters the root hair by osmosis?   Answer . . . .
Q.82 enters the leaf by refraction?   Answer . . . .
Q.83 translocates mainly food and sugars?   Answer . . . .
Q.84 is vascular tissue concerned with
      upward translocation of water?   Answer . . . .

Questions 85, 86, 87 and 88

    A. red blood cell
    B. blood plasma
    C. tissue fluid
    D. lymph
    E. white blood cell

Which of the above

Q.85   is a liquid found between muscle cells?   Answer . . . .

Q.86   is a nucleated cell capable of
             amoeboid movement?   Answer . . . .

Q.87   is a liquid deficient in protein, rich in
             lipids, and found in tubular vessels?   Answer . . . .

Q.88   is a non-nucleated cell concerned with
             oxygen transport?   Answer . . . .

Questions 89, 90, 91 and 92

    A. pulmonary vein
    B. pulmonary artery
    C. vena cava
    D. portal vein
    E. aorta

Which of the above components of the mammalian blood system

Q.89   is the largest artery in the body
             transporting blood from the heart?   Answer . . . .

Q.90   is a vein connecting two capillary networks?   Answer . . . .

Q.91   is a blood vessel carrying oxygenated
             blood from the lungs to the heart?   Answer . . . .

Q.92   is the blood vessel passing deoxygenated
             blood into the heart's right atrium?   Answer . . . .

16.7 Respiration

Questions 93, 94, 95 and 96

    A. $6CO_2 + 6H_2O + energy \rightarrow C_6H_{12}O_6 + 6O_2$
    B. $C_{12}H_{22}O_{11} + H_2O \rightarrow 2C_6H_{12}O_6$
    C. $C_6H_{12}O_6 + 6O_2 \rightarrow 6CO_2 + 6H_2O + energy$
    D. $C_6H_{12}O_6 \rightarrow 2CH_3.CHOH.COOH + energy$
    E. $C_6H_{12}O_6 \rightarrow 2C_2H_5OH + 2CO_2 + energy$

Which of the above chemical equations represents the main chemical changes occurring in

Q.93   anaerobic respiration in human skeletal
             muscle?   Answer . . . .

Q.94   anaerobic respiration in yeasts?   Answer . . . .

Q.95   photosynthesis?   Answer . . . .

Q.96   aerobic respiration?   Answer . . . .

Questions 97, 98, 99 and 100

    A. complemental air
    B. lung capacity
    C. tidal air
    D. vital capacity
    E. residual air

Choose the term from the above which means the volume of air

Q.97 which may be expelled by forced expiration
after a deep breath.   Answer . . . .

Q.98 which cannot be naturally expelled and is
that volume which remains in the lungs.   Answer . . . .

Q.99 moved in and out of the lungs with each
breath.   Answer . . . .

Q.100 which may be changed by a deep breath
without undue forcing.   Answer . . . .

Questions 101, 102, 103 and 104

    A. plasma membrane (plasmalemma)
    B. mitochondrion
    C. alveolus
    D. cytoplasm
    E. moist skin

Which of the above is

Q.101 part of the mammal respiratory tree?   Answer . . . .
Q.102 the gas exchange surface of an animal cell? Answer . . . .
Q.103 the respiratory organ of earthworms?   Answer . . . .
Q.104 the organelle of aerobic respiration?   Answer . . . .

Questions 105, 106, 107 and 108

    A. stomata
    B. epidermis
    C. chloroplasts
    D. root hairs
    E. lenticels

Which of the above

Q.105 has cell walls permeable to gases?   Answer . . . .
Q.106 is a location for oxygen production?   Answer . . . .
Q.107 functions for gas exchange in a woody
stem and root bark   Answer . . . .
Q.108 are pores for gas exchange in aerial
stems, flowers and leaves?   Answer . . . .

16.8 Growth

Questions 109, 110, 111 and 112

    A. anabolism
    B. cambium

C. vacuolation
D. elongation
E. mitosis

Which of the above

Q.109 is a process of cell and nucleus division to produce identical cells from a parent cell?   Answer ....

Q.110 is the name for all processes of synthesis of cell materials?   Answer ....

Q.111 is a process of forming fluid-filled cavities in cell cytoplasm?   Answer ....

Q.112 is a meristem that forms plant vascular tissue?   Answer ....

Questions 113, 114, 115 and 116

A. water supply
B. suitable temperature
C. oxygen supply
D. pituitary hormone
E. energy source

Which of the above

Q.113 is not an essential factor for plant growth?   Answer ....

Q.114 is present in broad-bean cotyledons before germination?   Answer ....

Q.115 is responsible for catabolism in seed germination?   Answer ....

Q.116 enters the seed by way of its micropyle?   Answer ....

Questions 117, 118, 119 and 120

A. auxins
B. determinate growth
C. endosperm
D. complete metamorphosis
E. DNA replication

Which of the above

Q.117 is an important process of synthesis occurring in a nucleus?   Answer ....

Q.118 are plant growth hormones?   Answer ....

Q.119 involves a change from larva to adult and ecdysis?   Answer ....

Q.120 is the storage tissue of many seeds?   Answer ....

Questions 121, 122, 123 and 124

A. growth disc
B. growth ring
C. growth factor
D. growth rate
E. growth movement

Which of the above

Q.121 is a response of a plant part to a directional
stimulus?   Answer . . . .

Q.122 is seen as the annual increase in girth of
woody stems and roots?   Answer . . . .

Q.123 is seen in certain parts of a long bone
which allows growth in length?   Answer . . . .

Q.124 is the amount of increase in unit time?   Answer . . . .

16.9   Homeostasis

Questions 125, 126, 127 and 128

    A. dialysis
    B. osmoregulation
    C. gas exchange
    D. osmosis
    E. deamination

Which of the above

Q.125 is the means by which animals maintain
the concentration of water in their bodies
at the correct level?   Answer . . . .

Q.126 is the means to separate proteins from
mineral ions using a semi-permeable
membrane?   Answer . . . .

Q.127 is a process occurring in mammal livers?   Answer . . . .

Q.128 is a process of water diffusion across a
selectively permeable membrane?   Answer . . . .

Questions 129, 130, 131 and 132

    A. sweat glands
    B. homoiothermic (endothermic)
    C. vasodilation
    D. hypothermia
    E. adipose tissue

Which of the above

Q.129 is a means of heat insulation in the mammal
body?   Answer . . . .

Q.130 will occur after strenuous physical
exercise in humans?   Answer . . . .

Q.131 can occur after prolonged immersion of
the human body in cold sea water?   Answer . . . .

Q.132 is a homeostatic condition found in birds
and mammals and not in fish and
amphibians?   Answer . . . .

Questions 133, 134, 135 and 136

    A. glycogen
    B. glucose
    C. insulin

D. pancreas
E. liver

Which of the above

Q.133 is a functional protein hormone?   Answer . . . .

Q.134 is a polysaccharide found mainly in
        liver and skeletal muscle?   Answer . . . .

Q.135 detoxifies foreign drugs, chemicals and
        ethanol?   Answer . . . .

Q.136 speeds up the uptake of glucose into
        skeletal muscle and liver cells?   Answer . . . .

Questions 137, 138, 139 and 140

A. skin
B. lungs
C. pancreas
D. kidneys
E. liver

Which of the above is the main homeostatic organ for

Q.137 body temperature control?   Answer . . . .
Q.138 plasma glucose concentration control?   Answer . . . .
Q.139 plasma water concentration control?   Answer . . . .
Q.140 plasma carbon dioxide concentration
        regulation?   Answer . . . .

16.10   Irritability

Questions 141, 142, 143 and 144

A. motor neurones
B. receptors
C. effectors
D. sensory neurones
E. intermediate neurones

Which of the above components of the nervous pathway

Q.141 are also called muscles, glands or cilia?   Answer . . . .

Q.142 are mainly located in the sense organs?   Answer . . . .

Q.143 are efferent neurones whose axons connect
        with effectors?   Answer . . . .

Q.144 are neurones connecting with the central
        nervous system?   Answer . . . .

Questions 145, 146, 147 and 148

A. rods
B. cones
C. cornea
D. lens
E. sclerotic

Which of the above component parts of the human eye

Q.145 requires vitamin A (retinol) for its
      efficient function?   Answer . . . .

Q.146 is opaque to light rays?   Answer . . . .

Q.147 produces the greatest refraction of
      light rays?   Answer . . . .

Q.148 is located between the aqueous humour
      and the vitreous humour?   Answer . . . .

Questions 149, 150, 151 and 152

    A. salivary gland
    B. endocrine gland
    C. neurone axon
    D. neurone synapse
    E. myelin sheath

Which of the above

Q.149 secretes a hormone?   Answer . . . .

Q.150 secretes enzymes?   Answer . . . .

Q.151 transmits electrochemical impulses?   Answer . . . .

Q.152 allows diffusion of chemical transmitter
      substances?   Answer . . . .

Questions 153, 154, 155 and 156

    A. adrenaline
    B. glucagon
    C. insulin
    D. knee-jerk reflex
    E. conditioned (learned) reflex

Which of the above

Q.153 is a reflex that does not involve an
      intermediate neurone?   Answer . . . .

Q.154 stimulates glucose uptake by liver cells? Answer . . . .

Q.155 can cause vasoconstriction?   Answer . . . .

Q.156 stimulates conversion of glycogen into   Answer . . . .
      glucose?

16.11 Support and Movement

Questions 157, 158, 159 and 160

    A. plasmolysis
    B. turgor (turgidity)
    C. network of vascular tissue
    D. central core of vascular tissue
    E. cylindrical ring of separate vascular bundles

Which of the above terms and descriptions concerning herbaceous dicotyledon flowering plants

Q.157 is the main factor in maintaining
      support in the unlignified parts of plants? Answer . . . .

Q.158 is the supporting tissue arrangement
      in roots?   Answer . . . .

Q.159 is the supporting tissue arrangement
resisting bending stresses?   Answer . . . .

Q.160 is the cause of wilting?   Answer . . . .

Questions 161, 162, 163 and 164

    A. flagellum
    B. cilia
    C. chitin
    D. collagen
    E. calcium salts

Which of the above

Q.161 is the means of movement in human
spermatozoa?   Answer . . . .

Q.162 is the main component of insect
exoskeletons?   Answer . . . .

Q.163 remains when a mammal bone is soaked
in hydrochloric acid?   Answer . . . .

Q.164 is a dietary component present in milk
and cheese?   Answer . . . .

Questions 165, 166, 167 and 168

    A. smooth (unstriped) muscle
    B. striated (striped) muscle
    C. end plate
    D. tendon
    E. ligament

Which of the above

Q.165 is the part of a motor neurone connecting
with skeletal muscle?   Answer . . . .

Q.166 is the main component of a flexor
(biceps) muscle?   Answer . . . .

Q.167 connects the extensor (triceps) muscle
to the bone?   Answer . . . .

Q.168 is the main component of the gut
pylorus sphincter?   Answer . . . .

Questions 169, 170, 171 and 172

    A. hyaline cartilage
    B. joint capsule
    C. antagonistic muscles
    D. involuntary muscles
    E. spongy bone

Which of the above

Q.169 produces new red blood cells?   Answer . . . .

Q.170 secretes synovial fluid from its lining
membrane?   Answer . . . .

Q.171 is strengthened with ligaments?   Answer . . . .

Q.172 describes the forearm muscles?   Answer . . . .

16.12   Reproduction

Questions 173, 174, 175 and 176

    A. spores
    B. perennation
    C. artificial propagation
    D. lateral buds
    E. runners

Which of the above

Q.173 are found in leaf axils?   Answer . . . .

Q.174 originate from leaf axils to form
        creeping overground stems?   Answer . . . .

Q.175 is a method of overwintering of trees
        and shrubs?   Answer . . . .

Q.176 are simple asexual unicellular
        reproductive units?   Answer . . . .

Questions 177, 178, 179 and 180

    A. plumule
    B. zygote
    C. ovule
    D. pollen grain
    E. pericarp and testa

Which of the above

Q.177 produces the flowering plant embryo
        by repeated mitosis?   Answer . . . .

Q.178 contains the male haploid nucleus?   Answer . . . .

Q.179 encloses the product of fusion of two
        gametes?   Answer . . . .

Q.180 is part of the flowering plant embryo?   Answer . . . .

Questions 181, 182, 183 and 184

    A. prostate gland
    B. oviducts
    C. testes
    D. uterus
    E. ovaries

Which of the above human reproductive parts

Q.181 secretes nutritive seminal fluid?   Answer . . . .

Q.182 is the normal location for implantation?   Answer . . . .

Q.183 is the normal location for fusion of
        gametes in human beings?   Answer . . . .

Q.184 secretes the sex hormones oestrogen and
        progesterone?   Answer . . . .

Questions 185, 186, 187 and 188

    A. placenta
    B. umbilical cord

C. uterus

D. vagina

E. amniotic sac

Which of the above human reproductive parts

Q.185 is a temporary organ which produces female sex hormones? Answer . . . .

Q.186 is a protective embryonic membrane and fluid? Answer . . . .

Q.187 undergoes periodic replacement of its epithelium accompanied by haemorrhage? Answer . . . .

Q.188 connects the human embryo with the placenta? Answer . . . .

16.13 Genetics and Evolution

Questions 189, 190, 191 and 192

A. homozygous

B. heterozygous

C. dominant

D. recessive

E. codominant

Which of the above terms

Q.189 designates a diploid nucleus with two identical alleles? Answer . . . .

Q.190 designates the character or trait seen in *both* the homozygote and the heterozygote? Answer . . . .

Q.191 designates the character or trait seen only in the homozygote? Answer . . . .

Q.192 designates the genetic condition in which alleles blend to produce an intermediate trait or character? Answer . . . .

Questions 193, 194, 195 and 196

A. *XX* chromosomes

B. *XY* chromosomes

C. mutation

D. fertilisation

E. autosomes

Which of the above

Q.193 are homologous chromosomes with the exception of sex chromosomes? Answer . . . .

Q.194 are homologous chromosomes associated with the female human? Answer . . . .

Q.195 involves a sudden change in the gene or chromosome transmitted to the offspring? Answer . . . .

Q.196 number a total of 44 in a normal human karyotype or karyogram? Answer . . . .

Questions 197, 198, 199 and 200

    A. mitosis
    B. meiosis
    C. continuous variation
    D. discontinuous variation
    E. non-heritable variation

Which of the above

Q.197 shows an average value among members
      of the same species?   Answer . . . .

Q.198 is a process producing cells with
      diploid nuclei?   Answer . . . .

Q.199 shows as sharp differences determined
      by one or two genes?   Answer . . . .

Q.200 is a process in which the nucleus
      chromosome number is halved?   Answer . . . .

Questions 201, 202, 203 and 204

    A. chemical evolution
    B. organic evolution
    C. natural selection
    D. analogous structures
    E. homologous structures

Which of the above

Q.201 is a process of genetic change among
      living organisms?   Answer . . . .

Q.202 is an example of the environment allowing
      certain genotypes to survive and reproduce? Answer . . . .

Q.203 have the same origin in a common ancestor
      but have developed different functions?   Answer . . . .

Q.204 have different origins but have
      developed similar functions?   Answer . . . .

16.14   Organisms and the Environment

Questions 205, 206, 207 and 208

    A. decomposers
    B. consumers
    C. producers
    D. energy forms
    E. chemical elements

Which of the above ecosystem components

Q.205 are autotrophic organisms?   Answer . . . .
Q.206 are not recycled?   Answer . . . .
Q.207 are recycled?   Answer . . . .
Q.208 are heterotrophic organisms attacking
      dead organic matter?   Answer . . . .

Questions 209, 210, 211 and 212

    A. population
    B. community
    C. habitat
    D. ecosystem
    E. niche

Which of the above

Q.209 is a natural unit of living and non-living parts interacting to produce a stable system?   Answer . . . .

Q.210 is the place where a living organism lives?   Answer . . . .

Q.211 is a group of individuals of the same species in a particular space?   Answer . . . .

Q.212 is the job or functional role of an organism in a community?   Answer . . . .

Questions 213, 214, 215 and 216

    A. saprophytism
    B. competition
    C. parasitism
    D. predation
    E. symbiosis

Which of the above describes

Q.213 birds eating worms?

Q.214 yeast growing on bruised windfall apples?   Answer . . . .

Q.215 fungus growing on leaves of a living potato plant?   Answer . . . .

Q.216 dead leaves decaying in soil?   Answer . . . .

    (NEA)

Questions 217, 218, 219 and 220

    A. digging over the soil
    B. rotating crops grown in an area
    C. adding farmyard manure
    D. lime spreading
    E. scattering artificial fertiliser

Which of the above agricultural and gardening practices

Q.217 quickly increases the plant nutrient level in the soil?   Answer . . . .

Q.218 helps control infectious diseases in plants?   Answer . . . .

Q.219 improves the water-retaining properties of a sandy soil?   Answer . . . .

Q.220 raises the pH value of the soil?   Answer . . . .

    (NISEC)

Questions 221, 222, 223 and 224

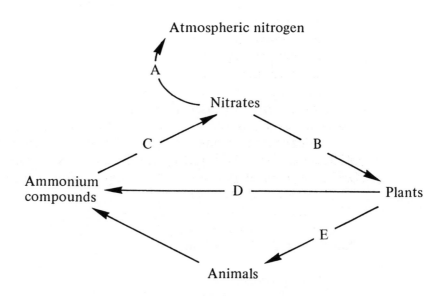

Which of the processes labelled A to E in the simplified diagram of the nitrogen cycle

Q.221   converts non-available nitrogen into a form that can be used by plants?   Answer . . . .

Q.222   leads to the production of amino acids?   Answer . . . .

Q.223   involves saprophytic fungi?   Answer . . . .

Q.224   decreases the amount of nitrogen available to most plants?   Answer . . . .

(NISEC)

Questions 225, 226, 227 and 228

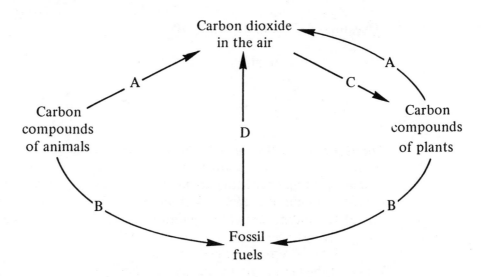

Which of the above processes labelled A to D in the simplified diagram of the carbon cycle indicates

Q.225   aerobic respiration?   Answer . . . .

Q.226   decay and decomposition?   Answer . . . .

Q.227   photosynthesis?   Answer . . . .

Q.228   chemical combustion?   Answer . . . .

(NISEC)

Questions 229, 230, 231 and 232

    A. scurvy
    B. Down's syndrome
    C. tuberculosis
    D. sickle-cell anaemia
    E. sugar diabetes

Which of the above diseases is caused by

Q.229 a gene mutation?   Answer . . . .

Q.230 nutrient deficiency?   Answer . . . .

Q.231 a species of bacteria?   Answer . . . .

Q.232 hormone deficiency?   Answer . . . .

Questions 233, 234, 235 and 236

    A. protozoa
    B. unicellular algae
    C. yeasts
    D. bacteria
    E. viruses

Which of the above are valuable organisms in biotechnology

Q.233 as sewage detritus feeders?   Answer . . . .

Q.234 as antibiotic producers?   Answer . . . .

Q.235 as a source of vitamins of the B group
       and as a fuel producer?   Answer . . . .

Q.236 as a source of vegetable protein, and
       also of oxygen in water purification? Answer . . . .

Questions 237, 238, 239 and 240

    A. manometer
    B. calliper
    C. quadrat
    D. pooter
    E. forceps

Which of the above would be correctly selected to

Q.237 pick up minute soft-bodied organisms? Answer . . . .

Q.238 pick up a pesticide-contaminated leaf? Answer . . . .

Q.239 measure the thickness of an ear lobe? Answer . . . .

Q.239 mark out an area of ground habitat? Answer . . . .

Questions 241, 242, 243 and 244

From the list below choose *one* which best fits each description in the table.
Write your answers in the spaces provided.

CONSUMER   POPULATION   BIOMASS
COMMUNITY   PRODUCER   HABITAT
COMPETITION   NUTRIENT CYCLE   DECOMPOSER

	Description	Term
Q.241	A group of individuals of the same species	
Q.242	An organism which breaks down leaf litter	
Q.243	Interacting populations within the same habitat	
Q.244	An organism which changes light energy into stored chemical energy	

(L & EAG)

Questions 245, 246, 247, 248 and 249

The diagram below shows part of the nitrogen cycle in a woodland. The labelled arrows represent different processes.

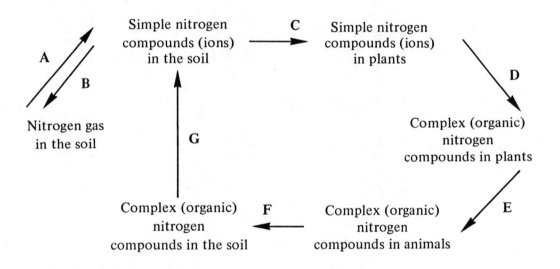

In the table below, write the name of a process occurring at each labelled arrow. Two lines have been completed for you.

	Arrow label	Process
Q.245	A	
	B	denitrification
Q.246	C	
	D	protein synthesis
Q.247	E	
Q.248	F	
Q.249	G	

(L & EAG)

Mark Scheme

16.1 Bioscience

1 C	2 D	3 A	4 B
5 E	6 C	7 D	8 B
9 D	10 A	11 A	12 C
13 D	14 C	15 A	16 B

16.2 Variety of Organisms

17 A	18 E	19 A	20 B
21 B	22 C	23 A	24 E
25 A	26 C	27 A	28 A
29 D	30 C	31 E	32 B

16.3 Cells, Tissues, Organs and Organisation

33 B	34 A	35 D	36 E
37 B	38 A	39 D	40 E
41 B	42 A	43 D	44 C
45 B	46 D	47 A	48 E

16.4 Plant Nutrition

49 D	50 B	51 C	52 B
53 A	54 C	55 E	56 B
57 B	58 D	59 E	60 C
61 C	62 A	63 D	64 B

16.5 Animal Nutrition

65 D	66 C	67 A	68 D
69 A	70 C	71 D	72 E
73 C	74 D	75 E	76 A

16.6 Transport

77 B	78 A	79 D	80 C
81 B	82 D	83 C	84 A
85 C	86 E	87 D	88 A
89 E	90 D	91 A	92 C

16.7 Respiration

93 D	94 E	95 A	96 C
97 D	98 E	99 C	100 A
101 C	102 A	103 E	104 B
105 D	106 C	107 E	108 A

16.8 Growth

109 E	110 A	111 C	112 B
113 D	114 E	115 C	116 A
117 E	118 A	119 D	120 C
121 E	122 B	123 A	124 D

Homeostasis

125 B	126 A	127 E	128 D
129 E	130 C	131 D	132 B
133 C	134 A	135 E	136 C
137 A	138 C	139 D	140 B

16.10 Irritability

141 C	142 B	143 A	144 D
145 A	146 E	147 C	148 D
149 B	150 A	151 C	152 D
153 D	154 C	155 A	156 B

16.11 Support and Movement

157 A	158 D	159 E	160 A
161 A	162 C	163 D	164 E
165 C	166 B	167 D	168 A
169 E	170 B	171 A	172 C

16.12 Reproduction

173 D	174 E	175 B	176 A
177 B	178 D	179 E	180 A
181 A	182 D	183 B	184 E
185 A	186 E	187 C	188 B

16.13 Genetics and Evolution

189 A	190 C	191 D	192 E
193 E	194 A	195 C	196 E
197 C	198 A	199 D	200 B
201 B	202 C	203 E	204 D

16.14 Organisms and the Environment

205 C	206 D	207 E	208 A
209 D	210 C	211 A	212 E
213 D	214 A	215 C	216 A
217 E	218 B	219 C	220 D

221 C	222 B	223 D	224 A
225 A	226 B	227 C	228 D or A (see No)
229 D	230 A	231 C	232 E
233 A	234 D	235 C	236 B
237 D	238 E	239 B	240 C
241 population	242 decomposer	243 competition	
244 producer	245 nitrification	246 absorption (active transport	
247 feeding (ingestion)	248 excretion, egestion	249 decomposition, decay	

Note: there are alternative answers to Q228, namely D – combustion of fossil fuels, and A – combustion of carbon compounds in plants and animals by bonfire or cremation.

Book List

The purpose of this list is to enable you to widen and enjoy your study of biology. Certain books should be purchased, and these are indicated by a '†'; those not marked '†' can be borrowed from a library.

General

† *Mastering Biology*, by O. F. G. Kilgour (Macmillan Education). A comprehensive theoretical and practical account to support this work book.

† *Multiple-choice Questions in Biology and Human Biology*, by O. F. G. Kilgour (Heinemann Educational Books). Over 1500 graded questions for practice.

† *Dictionary of Life Sciences*, by E. A. Martin (published by Macmillan Press). A concise and invaluable reference book.

Diversity of Living Organisms

Natural history is an enjoyable study and pastime. There is a wide selection of books to choose from, as indicated by the following:

'Discoverers' series (Methuen/Moonlight).
'Nature Notebook' series (Granada).
'Nature Notebooks', National Trust series (Reader's Digest).
'Observers' series (Warne).
'The Illustrated Book of . . .' series (Octopus).
'A Field Guide in Colour' series (Octopus).
'A Colour Guide to Familiar . . .' series (Octopus).
Diversity of Life, by Jenkins and Boyce (Macmillan Education). A book for GCE 'A' level studies.

Structure and Function of Living Organisms

The Cell Concept, by Kramer and Scott (Macmillan Education). A book for GCE 'A' level studies.

Biological Structures, by Krommenhuik, Sebus and Van Esch (John Murray). A collection of photomicrographs and electron micrographs with X-ray pictures of tissues, organs and systems.

Mastering Nutrition, by Kilgour (Macmillan Education). A study of nutrition from first principles.

Metabolism, Movement and Control, by Boyce and Jenkins (Macmillan Education). A book for GCE 'A' level studies.

Development of Organisms and Continuity of Life

Life on Earth, by David Attenborough (BBC Publications/Collins). A beautiful and well-illustrated book.

Heredity, Development and Evolution, by Birkett (Macmillan Education). A book for GCE 'A' level studies.

Relationships between Organisms and the Environment

Man and the Ecosystem, by Lloyd (Macmillan Education). A complete, advanced course.

Ecology Principles and Practice, by W. H. Dowdeswell (Heinemann).

Macmillan Guide to Britain's Nature Reserves, by J. Hywel-Davies and V. Thom (Macmillan Education). This book lists all the main nature reserves open to the public, where plant-life and animal-life studies can be enjoyed in country parks or by following nature walks and trails. Owners, managers or wardens of the sites can often provide information and literature.

All GCSE students in the UK should make contact with and visit their local nature reserve as part of their practical studies.

Information and Resources

Biological Materials and Apparatus

Local chemist shops or pharmacies can supply small amounts of certain chemical reagents and testing strips, and such apparatus as test tubes, or else they can order such items for you.

Griffin and George, Ealing Rd, Alperton, Wembley, Middlesex HA0 1HJ and Philip Harris Ltd, Lynn Lane, Shenstone, Staffs WS14 0EE are the two major suppliers of all biological requirements to schools and colleges.

Local opticians can supply hand lenses, magnifiers and small modestly priced microscopes and slides.

Information and Literature

Local libraries often stock a range of leaflets on health education, and can give you the names and addresses of your local naturalist, birdwatching or conservation group; such groups are well worth joining for invaluable field experience.

The *Health Education Council*, 78 New Oxford St, London WC1N 1AH issues a whole range of leaflets and booklets, but they can also be obtained from your local Health Education Officer whose name and address can be obtained from your library.

Population Concern, 231 Tottenham Court Rd, London W1A 9AE sells a valuable 'World Population Chart' showing where the 5 000 000 000 people of the world are living.

The *Nature Conservancy Council* has local branches throughout the UK. Their addresses are in the telephone directories or can be obtained from 19–20 Belgrave Square, London SW1X 8PY; posters and booklets can be purchased on a range of topics.

Natural history museums, and botanical and zoological gardens are located throughout the UK and are essential places for all students of biology to visit and to study at again and again throughout the year.

The *Institute of Biology*, 20 Queensbury Place, London SW7 2DZ provides careers information.

Local Authority Water Boards may stage local exhibitions on natural history and ecology close to their reservoirs (check with the address in the telephone directory.)

The *Forestry Commission* often has exhibition centres close to their woodlands. Locations can be found from the telephone directory.

The *Vegan Society*, 33–35 George St, Oxford OX1 2AY issues leaflets about the vegan diet, ecology and animal rights.

Dr Hadwen Trust for Humane Research, 46 Kings Road, Hitchin, Herts SG5 1RD issues leaflets on the alternative use of *cell culture* methods and *biotechnology* instead of the use of animals (a) generally in the laboratory, (b) as sources for testing chemicals, (c) for the diagnosis of disease and (d) for the production of insulin.

The *Friends of the Earth*, 377 City Road, London EC1V 1NA issue literature on a range of ecological topics, which makes us aware of the duty of the human species to care for the Earth and life upon it, and to ensure that we leave it in a fit state for future generations of humans to inhabit.

The *Royal Society for the Protection of Birds*, The Lodge, Sandy, Bedfordshire SG19 2DL.

The *Royal Society for the Prevention of Cruelty to Animals*, The Causeway, Horsham, West Sussex RH12 1HG.

The *Royal Society for Nature Conservation*, The Green, Nettleham, Lincoln LN2 2NR.

Index*

*Page numbers in *italics* refer to illustrations. Alternative terms are given in brackets. '*Et seq.*' means 'and the following pages'.

digestion, 85 *et seq.*, 86, 94
digits, *189*
diploid chromosome number, 205, 211, 227, 229
disaccharides, 5, 86
discontinuous variation, 229
disease, 249, 250 *et seq.*, 278, 279
    defence against, 249, 259, 260
disinfection, 250, 280
dispersal of fruits and seeds, 212, 224, 226
dissection, 164, 280
dodder (parasite), *248*
dog, 22, *84*, 242
dominance, 229 *et seq.*, 238
dormancy, 148
dorsal, 36
drying methods, *277*
Dryopteris, *23*
ductless (endocrine) glands, *174*
duodenum, 86, *87*
dwarfism, 23

ear, 34, 35, 172, 177 *et seq.*, *178*, 184
    drum, *178*
earthworm, 28, *31*, 120, *188*, 252, 261, 284
ecdysis, 32, *146*, 147, 188
echinoderms, 28, *32*
ecology, 245 *et seq.*, 250, 307
    practical work, 250 *et seq.*
ecosystem, 245 *et seq.*, 261
ectoparasites, 247
ectoplasm, *29*
ectotherms (poikilotherms), 161 *et seq.*, 242
effectors, 159, 160, 172, 174, 191
effort, *191*, *192*
egg, 28, 32, *146*
elastin, 79
elbow joint, *190*
electromagnetic spectrum, *3*
electron micrographs, 71, 283
electron microscopes, 54, 283
embryo, 30, 142, *147*, *211*, 213, *215*, 224, 248
emulsification, 9, 86
enamel (tooth), *84*
endocrine system, 172, 173, *174*
endodermis (root), *99*
endoparasites, 247
endoplasmic reticulum, *50*, 60, 97, 98
endoskeleton, 188, *191*, 199
endosperm, *142*, *143*, 211
endotherms (homoiotherms), 161, 169, 242
energy, 3, 21, 63, 78, 97, 118, 140, 245, 246, *247*, 274
    body needs, 80, 125
    flow, 97, 140, 246, *247*
    food, from, 63, 118, 245
    sun (solar), 97, 99, 245, *247*, 258, 262
environment, external, 162, 172, 232, 245 *et seq.*, 307 *et seq.*
    internal, 159, 172, 245
enzymes, 4, 5, 6, 12, 14, 79, 85 *et seq.*, 143, 242, 265
    experiments with, 89 *et seq.*
epicotyl, *142*, *143*
epidermis, 52, *53*, *66*, *67*, 99, *100*, *103*, *144*, *162*, 212, *213*, 214
epididymis, 212, 213

epigeal germination, 143, *144*, 153
epiglottis, *87*, *122*
epithelium, 60, 85, *88*, 98, 101, 120
erepsin, 86
essential amino acids, 79
estimation, 274
eukaryotes, 24, 49, 50, 51, 227
eustachian tube, *178*
eutherians, 22, 28, 35
evaporation, 99, *100*, 116, 161, 247, 277
evolution, 232 *et seq.*, 236, 242, 306 *et seq.*
excretion, 22, 140, 163 *et seq.*, 168, 171, *189*, *215*
exocrine glands, 172, 191
exoskeleton, 28, 45, 188, *190*, *191*, *192*, 199
experimental work, 7 *et seq.*, 36 *et seq.*, 54 *et seq.*, 68 *et seq.*, 89 *et seq.*, 104 *et seq.*, 127 *et seq.*, 148 *et seq.*, 164, 179, 193, 215, 233, 250 *et seq.*, 266 *et seq.*
expiration, 120, 123, 128, 137
extensor muscle, 192
external respiration (breathing, ventilation), 120 *et seq.*
extracellular enzymes, 4, 85
eye, *31*, *32*, *33*, *34*, *35*, *82*, *83*, *146*, 147, 172, 176 *et seq.*, *177*, *183*, 184
eyepiece, microscope, 54, *281*

F (filial) generation, 230 *et seq.*
faeces, *87*, *248*
fallopian tubes (oviducts), 213, *214*
fat depots, 60, 79, 161, *162*
fats, *see* lipids
fat-soluble vitamins *A*, *D*, *E* and *K*, 79 *et seq.*
fatty (alkanoic) acids, 5, 86, 88
feathers, 28, *35*, 161
feedback, 1, 59, 173, *175*, 185
feeding methods, 82
Fehling's test, 9, 275
femur, 189, 190, *196*, 197
ferns (Pteridophyta), 23, *26*, 208, 209, *210*, 211, 215
fertilisation
    animals, 213, 218, 229, 242
    plants, *211*, *224*, 226, 229, 242
fibrin, 102
fibula, *189*, 190
filament, 210
filtration, 97, 159, *276*
fins, 28, *33*, *34*
first aid, 267
fish, *33*, 36, 120, *121*, 265
flagellum, 24, *25*, 54, 172, 188
flaming, 283
flatworms (Platyhelminthes), 28, *29*
flexor muscle, *192*
flow diagrams, 283, *284*
flower
    pollination, *211*
    structure, *27*, 205, 210, *211*, 215, *216*, *221*, *224*, 226
flowering plants, *see* angiosperms
fluoride, 80, 81
focusing
    eye, 176, *177*
    microscopes, 54 *et seq.*
foetus, 213, 215

foliage leaf, *see* leaf
folic acid, 78, 79
follicle, 213
food, 79 *et seq.*, 161, 162, 172
    additives, 81
    chain, *246*
    composition, 79 *et seq.*, 92
    energy value of, 79, 118
    poisoning, 250, 279
    tests, 9, 89 *et seq.*, 150
    webs, *246*, *258*, 265
forceps, 280
forelimb, *189*, *192*, *194*, 198, 200
foreskin, 214
fossils, 242, 247
fovea, *177*
fragmentation, 205
free nerve endings, 174, 176
freshwater pond, 245
frog, 28, *34*, 120, 145, *146*, 171
fructose, 5, 6, 9
fruit, 23, *27*, *143*, 211, *212*, 215, *220*, 224
fuels, 247, 248, 262
Funaria, 23, 24, *25*
fungi, 23, *24*, 124, 215, 245, 249, 250, 252

galactose, 5
gall bladder, 86, *87*, *163*
gamete, 205, 208, *211*, 230, 239
gametogenesis, 212
gametophyte, 208
gamma rays, *3*, 250
gas analysis, 74, 75, 78, 128
gas burners, 268, 269
gas exchange, 67, 68, 97, 120 *et seq.*, *123*
gastric juice, 86 *et seq.*, 249
genes, 227 *et seq.*, *232*
genetic engineering, 232, 249
genetics, 227 *et seq.*, 306 *et seq.*
    experiments, 233
genotype, 228, *230*, 236, 239
geotropism, 149, *150*, 185
germination, 142, 153
gestation, 214
gigantism, 178
gills, *33*, *34*, 82, 120, *121*, 147
glands, 52, 159, 174, 191
glans penis, *214*
glasswork, 267 *et seq.*
glenoid cavity, 189
gliding joint, *190*
glomerulus, 159, *161*
glucagon, 163 *et seq.*, *175*, 178, 185
glucose, 5, 6, 9, 63, 79, 86, 88, 118, 129, 163 *et seq.*, 169, *175*, 245, 262, 278
glycerol (propane triol), 86, 88, *282*
glycogen, 51, *57*, 62, 79, 88, 163 *et seq.*, *175*, 200
goitre, 81
golgi body, *50*
graafian follicle, 213, *215*
grafting, 206, *208*
graphs, 2, *17*, 71, *121*, *126*, *139*, 148, *154*, *171*, *186*, *263*, *264*, 266, 287, *289*
grass, 23, *221*
grasshopper, *32*, 145, *146*
greenhouse effect, 262

grey matter, *176*
growth, 21, 79, 140 *et seq.*, 185, 205, 227, 299 *et seq.*
   curve graph, *148*, 154
   experiments, 148 *et seq.*
   hormones, 148, 178, 185, 249
   rings, 145
guanine, 4, 5
guard cells, 66, *67*
gum (tooth), *84*
gut (alimentary canal), 51, 85, *87*, *93*, 96, 101, *103*, *189*, 191, 197, 200
gymnosperms, 23, *26*
gynoecium, 210, *211*

habitat, 36, 245, 248
haemoglobin, 101
haemolysis, 101
haemophilia, 236, *241*
hair, 28, *33*, *35*, 45, 161, *162*, 178, 242
halophyte, 64
haploid chromosome number, 208, 211, 227, 229
haustoria, 248
hazard symbols, 274, *275*
hearing sense, 172, 176
heart, 101, *103*, 104, *112*, 114, 116, *122*
heat, 3, 99, 118 *et seq.*, 125, *129*, 136, 161, 172, 250, 259, 262, 269
   balance in mammals, 162 *et seq.*
hepatic artery and vein, *103*, *163*
hepatic portal vein, *103*, *163*, 185
herbicides, 99, 248
herbivores, *31*, *32*, *33*, *35*, 79, 85, *93*, 245, 260
heredity, 148, 229 *et seq.*
hermaphrodite, 210
heterotrophic nutrition, 23, 44, 45, 79 *et seq.*, 245
heterozygotes, *230*, 232, 240
hibernation, 148
Hill (light) reaction, 64, 65, 76
hilum, 143
hindlimb, *189*
hinge joint, 189, 190
hip joint, *189*, 190
holophytic nutrition, 23
homeostasis, 21, 51, 102, 159 *et seq.*, 169, 172, 173, 185, 301 *et seq.*
   experiments, 164 *et seq.*
homoiothermic (endothermic), 161, 169, 242
homologous chromosomes, 227, 229
homozygous, 229, *230*, 232
hormones, 5, 79, 85, 148, 160, 173 *et seq.*, 178 *et seq.*, 185, 232
horse chestnut twig, *207*
host, 247, *248*
housefly, 21, 28, *83*
humerus, 189, *190*
humidity, 271
humus, 252, 259, 261
Hydra, 28
hydrogen, 65
   carbonate indicator, 9, *128*
   peroxide, 10, 275
hydrolysis, 7, 85, 118, 143
hydrophyte, 64
hydrostatic skeleton, 188, 199
hygrometers, 271

hygroscopic, 271
hyperthermia, 161
hypervitaminosis, 81
hypha, *24*, 45, 199, 250
hypocotyl, *142*
hypogeal germination, 143, *144*, 153
hypothalamus, 159, 173, *176*
hypothermia, 161

identification keys, 36, 38, 39, 41, 215
ileum, *87*
imago, 145, *146*
immovable joint, 190
immunity, 249, 260
implantation, 212, 223
incisor teeth, *84*
incomplete dominance (codominance), 230, *231*
incomplete metamorphosis, 145, *146*
incus (anvil), *178*
indeterminate growth, 147
infra-red rays, *3*, 64, 270
ingestion, 82 *et seq.*; 240
inner ear, *178*
inorganic compounds, 4, 63, 252
insecticide, 249
insects, 28, 32, 42, 45, 120, *122*, *145*, 188, 192, 210, 224, 252
insertion muscle, *192*
inspiration (inhalation), 120, 123, 128, 137
insulin, 88, 163 *et seq.*, *175*, 178, 185, 249
intercellular spaces and fluid, 98, *99*, *100*
intercostal muscles, 120, *122*, 123
intermittent growth, 147
internal environment, 159
internal respiration, 118
interphase, 227
intestinal juice, 86 *et seq.*, *88*
intracellular enzymes, 4, 64, 65, 98
intracellular transport, 98
invertebrates, 28, 205
involuntary muscle, 189, 200
iodine, 9, 80, 186, 275, 282
   test for starch 69, 150
ionising radiation, *3*
ions, 4, 98, 99
iris, *177*
irritability, 21, 172 *et seq.*, 205, 302 *et seq.*
   experiments 179
islets of Langerhans, 163, 174
isotopes, 4

jaw, *83*, *84*, 147
joints, 188, *190 et seq.*, 198, 200
joule, 3, 64, 274
jugular vein, *103*

kangaroo, 28
keratin, 5, 79
keratomalacia, 81
keys (identification), 3, 38, 39, 41, 215
kidney, *103*, 159 *et seq.*, *160*, 163, 167, 170, 171, *214*
kilogram, 1
kilojoule, 3, 79, 118, 119, 274
kilopascal, 3, 98, 104
kingdom, 21, 23, 28
klinostat (clinostat), 149, *150*
knee joint, *190*
koala bear, 28
kwashiorkor, 81

laboratory techniques, 267 *et seq.*
labour (birth), 214
lactation, 80, 127, 214
lacteal vessel, *88*, 101
lactic acid, 118, 121, 129, 134, 139
lactose, 5, 86
large intestine, 86 *et seq.*, *87*
larva, 145, *146*
larynx, 122
lateral root, *27*
leaf, 23, *25*, *27*, 65, 78, 98, 104, 169, 207
   fall, *207*
   structure, 27, 66, 75, 76, 99, *100*
   suction, 98, 99, 105, *106*
Lecidea, 23
legume, 226
length measurement, 147, 148, 271, *272*
lens, 54, 176, *177*
lenticels, 99, *124*, 169, *207*
leucocytes (white blood cells), *102*, 117
levers in limbs, *191*, *192*
lichens, 23, 247
life characteristics, 21
life process, 205
ligaments, 189, *190*, 198
light, 3, 63 *et seq.*, 69, 70, 71, 77, 124, 125, *126*, 143, 148, 172, 176, 245
   microscopes, 54
   spectrum, 3, 64
lignin, 52, 188
limbs, 189 *et seq.*, *191*, *192*, 193
lime water, 9, *128*
liming, 261
limiting factors, 70, 71, 77
line graphs, 287
line transects, 250
lipase, 85, 86
lipids (oils and fats), 5, 9, *57*, 79, 80, 94, 119, 150, 161
litre, 1, 273
liver, 86, *87*, 88, *103*, 119, 162, 163, *249*, 260
liverworts, 23, 211
living organisms, 21
lizard, *28*
load, *191*, *192*, 193
locomotion, 21, 172, 190, 191 *et seq.*
locus (gene), 227
longitudinal muscle, *88*, *189*
long sight, *177*
lung, *103*, *122*, 128, 163
lymphatic system, *88*, 101, *102*, 117, 249, 260
lysosomes, 85

magnesium 65
magnification, 54, 281
magnifiers, 280
maize, *70*, *143*
malaria, 250
malleus (hammer), *178*
malnutrition, 81 *et seq.*
maltase, 86
maltose, 5, 9, 86
mammals, 22, 28, *35*, 45, 161, 168, 212, 232, 242
mammary glands, 28, 45, 178, 214
manometer, 98, 105, *274*
marasmus, 81
mass spectrometer, 4

plasma
    blood, of, 88, 102, *123*
    membrane, 49, *50*, *66*, *101*, 120, 141, 188, *213*
plasmolysis, 101
plastids, 22
platelets, *102*
Platypus, 28
pleura, *122*, 123
plexus, *176*
plum fruit, *212*
plumule, *142*, *143*
pod, 221, 226
poikilothermic (ectothermic), 161
poisons, 6, 105, 106, 271
pollen, 210, *211*
pollination, 210, *211*, *221*, *224*, 226
pollution, 248, 262
polysaccharide, 5, 7
pooter, *280*
population, 205, 229, 232, 245, 257, 263
pores, 97, 99
Porifera (sponges), 28
pork, *248*
potassium, 65
    hydroxide, 9, 75, 78, 128
potato, *74*, *206*, 282
potometer, 105, *106*, *107*
predator–prey relationship, 257, *263*
pregnancy, 80, 127
premolar teeth, *84*
pressure, 3, 104, 105, 123, 172, 274
primrose, *224*
privet, *66*
proboscis, *82*, *83*, 224
producer, 22, *25*, *26*, *27*, 245 *et seq.*, *247*, 258, 261, 265
proglottid, *30*, 248
prokaryotes, 24, 49, 50
prophase, *228*
prostate gland, 212, *213*, *214*
proteinase (protease), 85
proteins, 5, 7, 9, 65, 79, 86, 95, 119, 150, 161, 227, 228, 245, 249, 250, 257, 264
prothallus, 208, 209, *210*
Protophyta (Protista), 21
prototherians, 28
Protozoa, 21, 28, *29*, 249, 250, 252
pseudopodium, *29*, 82
psittacosis, 278, 279
Pteridophyta, 23, *26*
pulp (tooth), *84*
pulse rate, 182, *186*
Punnett square, *230 et seq.*
pupa, 145, *146*
pupil, *177*
pyramid
    food chain, in, 246
    kidney, of, *160*
pyrenoid, *25*
pyrogallol, alkaline, 9, 75, 78, 128, 275

quadrat, 250
qualitative variables, 1, 270
qualitative variations, 229, 233
quantitative variables, 229, 233, 270
quantitative variations, 229

rabbit, *35*, *87*
radiation, *3*, 4, 63, 64, 78, 97, 161, 242, 250
radicle, *142*, *143*, *149*
radioactivity, *3*, 4
radio waves, *3*
radius, *189*, *192*
radula, *31*, 82
ratio calculations, 2, 286
rats, 279
razor, 282
receptacle, 210, *224*
receptors, 159, 172, *173*
recessive traits, 229 *et seq.*, 231, 238, 240
rectum, *87*, *160*, *214*
recycling chemical elements, *247*
red blood cells, 60, *101*, *102*, *123*, 190
reducing sugars, 9
reflex action and arc, 173, *174*, 179
refraction, 99
relay neurone, 173, *174*
renal artery and vein, *103, 160, 161*
rennin, 86
replication, DNA, 140, 227, 250
reproduction, 21, 205 *et seq.*, 250 *et seq.*, 305 *et seq.*
    experiments, 215
reproductive systems, 51, *214*, 223
reptiles, 28, *34*, 242
respiration, 21, 51, 118 *et seq.*, 126, 132, 168, 190, 205, 247, 298 *et seq.*
    experiments, 127 *et seq.*
respirometers, *127*
response, 172, 182, 187, 205
retina, 176, *177*, 179
rhizoid, 23, *25*, *206*, 210
rhizome, *26*, 205
rib cage, 120, *122*, 123, 190
ribonucleic acid (RNA), 5, 250, 264
ribose, 3, 5, 9
ribosome, 24, 50, 264
rickets, 81
rods (eye), 176
root, 23, *27*, 53, *58*, 98, *99*, *144*, *153*, 172
    hair, 53, *99*, 100, 124, *144*, *153*, 169, 261
    nodules, 247
    pressure, 98, *105*
rose fruit, *212*
round window, *178*
roundworms, 28, 31

saccule, 176, *178*
salivary glands, 83, 86, 89
salts, 64, 159
sampling technique, *251*
saprophyte, 45, 250
scale leaf, 65, *206*
scapula, *192*
scent, 211, 226
scientific method, 1
scion, 208
scissors, *280*
sclerenchyma tissue, 52, 53, 60, 188, 199
sclerotic (sclera), *177*
scolex, 30, 248
scrotum, *214*
scurvy, 81
sea anemone, 28, *29*

sebaceous glands, *162*
secondary growth (thickening), 142, *145*, 150, 155, 188, 193, 199
secondary sexual characteristics, 178
section cutting, 54, *160*, *176*, *177*, *183*, 215, 224, *282*
seed
    formation, 23, *27*, *211*, 212, 224
    germination, *142*, 153, 156
    structure, *142*, *143*, 156, 231
seedless plants, 209
segmentation, 28, *31*
self-fertilisation, *230*
self-pollination, 210
semicircular canals, 176, *178*
seminal fluid, 212
seminal vesicles, *214*
seminiferous tubules, 212, 215
sensory cells, *162*, *174*, 178, 179
sensory neurone, *173*, *174*, 176
sepals, 65, 210, *224*
septum, *24*, *189*
serum, 102
sewage, 250
sex
    chromosomes, 225, 228
    determination, 127, *232*
    hormones, 176, 178
    linkage, 236, *241*
sexual reproduction, 205, 209 *et seq.*, 212, 223, 250
sheep, 84
shell, 28, *31*, 188
shivering, 162
short sight, *177*
shrimp, 28, *32*
siderosis, 81
sieve plate and tube, 52
skeletal muscle, 51, 118, 162, 189, 191, 193
skeleton, 51, 190, 197
skin, 5, 28, *33*, 34, *82*, 120, 161 *et seq.*, *162*, 178, 179, 225, 242, *249*, 260
skull, 93, 189, 193
small intestine, 86 *et seq.*, *87*, *88*
smooth muscle, 189, 191, 200
snail, *28*, 252
soda lime, 9, *127*, 275
soil, 99, 100, 124, 245, 251 *et seq.*, 257, 260, 261 *et seq.*
solar energy, 67
solution, 97, 265
    strength/concentration, 1, 6, 279
solvents, 97, 278, 279
sound, 172
speciation, 232, 242
species, 36, 127, 232, 245, 248, 257
spectacle lens, *177*, 183
Spermatophyta, 23, *27*, 212, *213*
spermatozoa, 212, *213*, 232
spermatozoids, 209, *210*
sphincter muscle, 30, 191
spider, 28, 33
spinal cord, *174*, 176
spiracles, *32*, *33*, 120, *122*, *146*
Spirogyra, 23, *25*, 42, 64, *209*
spirometer, 127
sponges, 28, *29*

xanthophyll, 64
X chromosome, 225, 228, 232, 236
X-rays, *3*, 250
xylem, 52, *53*, 60, 98, *99*, *100*, *145*, 199, *248*

Y chromosome, 225, 228, 232
yeast, *57*, 118, 119, *129*, *133*
yellow spot (fovea), *177*
yolk cytoplasm, *213*

yolk sac, *147*, *215*

zygospores, *209*
zygote, 205, 208, *211*, 213, 223, 224, 229